MAGIC, MATHEMATICS, AND PLAYING CARDS

MAGIC, MATHEMATICS, AND PLAYING CARDS

Jorge Nuno Silva
University of Lisbon, Portugal

Pedro J Freitas
University of Lisbon, Portugal

Alexandre Silva
Colégio de S José do Ramalhão, Portugal

Tiago Hirth
University of Lisbon, Portugal

NEW JERSEY • LONDON • SINGAPORE • BEIJING • SHANGHAI • TAIPEI • CHENNAI

Published by

World Scientific Publishing Co. Pte. Ltd.

5 Toh Tuck Link, Singapore 596224

USA office: 27 Warren Street, Suite 401-402, Hackensack, NJ 07601

UK office: 57 Shelton Street, Covent Garden, London WC2H 9HE

Library of Congress Control Number: 2025011611

British Library Cataloguing-in-Publication Data
A catalogue record for this book is available from the British Library.

MAGIC, MATHEMATICS, AND PLAYING CARDS

ISBN 978-981-98-0742-0 (hardcover)
ISBN 978-981-98-0816-8 (paperback)
ISBN 978-981-98-0743-7 (ebook for institutions)
ISBN 978-981-98-0744-4 (ebook for individuals)

For any available supplementary material, please visit
https://www.worldscientific.com/worldscibooks/10.1142/14165#t=suppl

Typeset by Stallion Press
Email: enquiries@stallionpress.com

To the Mathematical Circus, *a place of magic,*
to its predecessors, the Matemágicos Silva,
and to all the friends who helped test these effects,
we dedicate this book.

— *The authors*

Foreword

Magic, Mathematics, and Playing Cards

Magic is an age-old human activity; it exists since the dawn of humanity. One can find very old records of this art and of its uses in marvelling, impressing, or scaring people. In certain caves in North Africa and France, there exist prehistoric paintings from the Palaeolithic age, some more than 20,000 years old, which, according to experts, depict magic scenes.

We find another example in the Bible, in the book of Exodus. As Moses and his brother Aaron ask to see the Pharaoh, to demand freedom for their people, the Pharaoh asks them to perform a miracle as proof of their divine mission. Aaron then casts his staff on the ground and turns it into a snake. The Pharaoh, however, sends for his own Magicians who can also perform this trick, turning their staffs into snakes. Therefore, the Pharaoh doesn't let the Israeli people leave.

The same episode appears, with slightly different details, in the Koran, the sacred Muslim book. The Torah, the sacred book of Jews, also tells this story, with an added detail: Aaron takes the snake back in his hand, holding it by the tail, and turns it back into a staff.

For centuries, magic has been mostly at the service of people in power and of religions. Only recently has it been regarded as entertainment.

Mathematics is also a very ancient and universal subject. In all civilizations that left written records, there are references to mathematical activities and applications. It probably appeared, in an informal form, with the first humans and began to develop when mankind started to be organized in more complex ways. In the records of undeveloped societies, one can find signs of the use of mathematics. In prehistoric paintings and engravings, even among the oldest ones, in Lascaux, France, there are geometric forms and representations. From those times onward, these records become more and more frequent, and mathematics becomes a part of the culture of all human communities, growing exponentially and becoming the foundation for science and technology.

Playing cards are more recent. They were probably invented in China, and the first records of their use dating back to the 9th century. From China, they travelled to Persia and India. The constitution of the deck varies over time and by region. They probably reached Egypt in the 12th century, and it was probably from there that they reached Europe. Card games quickly expanded and became more popular, with decks and their symbology varying from country to country and even from city to city.

The use of cards for magic tricks probably followed quickly. The painting "The Conjurer", by Hieronymus Bosch, painted around 1500, already shows a magician in action. The first book on magic tricks was published in the late 16th century, and afterwards, card magic became more and more popular, reaching its peak in the last century.

The four authors of this book had the good idea of collecting, and offering us, more than 90 tricks (some with variants), where these three elements — Magic, Mathematics, and Playing cards — come together. All four of them are accomplished Magicians and handle playing cards with great ease. I have been a happy witness to this in some occasions.

This shows in the way the book is organized. It starts with easier tricks, moving on progressively to the harder ones, with great detail in explanation, following several steps: The presentation of the trick, the tricky details, the reason why it works, the mathematics behind it.

One might think that mathematics (a serious subject) doesn't really go well with magic or conjuring (which are forms of entertainment). This is false, of course, as you can see by reading this book.

The connection between magic and mathematics can be intriguing, thrilling, and spectacular.

With this book in hand, we, amateur Magicians, can

1. choose the tricks that better adjust to our abilities and style,
2. have fun performing them to family and friends (after some practice, of course),
3. go further and learn the mathematics behind them, following the explanations.

Let's have fun!

José Paulo Viana
Retired Mathematics teacher and author of books
on mathematical problems

Preface

In Gardner (2020), Martin Gardner quotes the following dialog from the short story *Mr. Know-All* by Somerset Maugham:

> "Do you like card tricks?"
> "No, I hate card tricks," I answered.
> "Well, I'll just show you this one."
> He showed me three. Then I said I would go down to the dining room and get my seat at table.

In the beginning of Chapter 1 of Diaconis and Graham (2015), the authors reinforce this idea:

> Most mathematical tricks make for poor magic and in fact have very little mathematics in them. The phrase "mathematical card trick" conjures up visions of endless dealing into piles and audience members wondering how long they will have to sit politely.

Despite all these warnings, and being well aware of the risk, we decided to write a book on mathematical magic. We have collected both classic and recent tricks, and a few original ones. We also propose new effects based on old principles, which should serve as an invitation to the reader to create their own.

We would like to emphasize that all the tricks we present in this book are mathematical: They depend solely on the combinatorial structure of the deck of cards, or on numerical principles, or on well-thought-out ways of shuffling, or on ingeniously created orders, or on some other mathematical concepts. In none of them is it necessary to use prestidigitation — although the combination of this skill with mathematics principles results in magnificent effects, such as "Henry Christ's Fabulous Four Aces".

It seems to us that, despite the dangers contained in the warnings with which we opened this preface, there are tricks that are not only mathematically interesting but also scenically attractive. We have tried to maintain both points of view. Thus, we have tried to present, in a simple but complete way, the mathematics behind the effects, as well as describe the conversation that needs to take place, the story that needs to be told, in order to make each trick an interesting experience for the audience.

Our description of the card tricks follows our routine presentations. For each one, we have different roles for the Magician (one of us), the Volunteer (from the audience), and the Helper (another person from our team). After going through the first few, the reader will, we hope, get familiar with our method.

For each activity, we first show the audience point of view (*Effect*), then we move to the recipe followed by the mathemagicians (*Method*), and, finally, we give a mathematical explanation for the trick to work smoothly (*Mathematics*).

As the structure of the performances is absolutely mathematical, everyone can replicate our activities. No special skill is required!

Contents

Foreword — vii

Preface — xi

1. Nice and Easy — 1

Quick tap — 1
A game with colors — 2
Klein — 3
One card out of the pocket — 5
Four Kings at the inn — 6
The poison card — 7
A square of cards — 8
The Belchou Aces — 8
The four nines — 9
The four eights — 10
Two Volunteers — 10
Two Fibonacci cards — 11
Lucky piles — 12
Four piles — 13
Three card guess — 14
The magical sum — 15
Out of this world — 17

2. Counting the Cards   **19**

The date 19
Coins in the pocket 20
Colorful piles 22
Buried treasure 24
Spelling bee 26
Green's "Love" 27
Three cards and three numbers 28
The magic of Manhattan 30
Numerology up and down 30
Remembering the future 32
Kruskal 33
Kraus 34
The keystone card discovery 35
Sum and difference 36
The fingerprint 38
Two cuts 40
The guessing Jokers 41
To impress the ladies 43
Five piles 45
Penelope's coincidence 47
The Elmsley coincidence 48

3. Shuffle and Deal   **53**

Rainman 53
Australian 54
Three Australians 54
Australian with company 56
Lost and found 57
Soul mate 58
The whispering Joker 60
Looking for the Joker 63
31 cards 65
Daisy's socks 67
Ace, two, and three 69

Luísa's twist 72
Liberty, fraternity, and equality 74
Five/three 75
The lady on the train 76
The Hummer equality 78
The Hummer Sum 79
The Hummer flush 80

4. Secret Codes 85

Out of the pocket 85
Turning rectangles 86
Queens and Kings 87
Tell me what I know 89
The programmed deck 90
The footsteps of the thief 92
Erdős 94
Three coins 96
Turn one 97
Pacioli's two rows 101
Three times seven, 21 104
Three times nine, 27 105
Three rows — but *not that trick!* 111
Memory of colors 115
One fortune serves eight 118
Cheney five-card trick 121

5. Gilbreath Galore 133

One black, one red 133
The swapped pair 135
Broken Gilbreath 136
A pile, a number, and a color 137
Fourteen cards 138
Seven days a week 140
The phone number 141
The four magic cards 142
The Diaconis poker 144

6.   Order in the Ranks!   147

You choose 147
Two sequences 149
Unfortunate 57 150
How many cards have passed? 152
Magical Tic-Tac-Toe 153
The six piles 156
Automatic counting 159
Sequeira divination 161
Sequeira spelling 162
Two cards indicate another 163
Where is the card? 164
Three in one 166
Joining by suits 166
Where are the red ones? 169

7.   Short Biographies   173

Alex Elmsley 173
Bob Hummer 173
Dai Vernon 174
Lennart Green 174
Martin Gardner 175
Persi Diaconis 177

Appendix   179

Ways of shuffling 179
   False 179
   Dovetail aka riffle 180
   Reverse aka in-and-out 180
   Rosetta 181
   Cut 181
   Monge 181
   Milk aka Klondike 182
   Australian aka down-and-under 182
   Hummer 183

Prepared decks 184
    Suits 184
    Progression 184
    Eight Kings 185
Positional systems 185
Modular arithmetic 186

Bibliography   189

Index   191

Nice and Easy

In this chapter, we present tricks in which both the execution and the mathematics involved are very simple — but which give rise to nice effects nonetheless. Some of these can even be learned by children, some of whom will even be able to understand the principles that make the tricks work. The group *Matemágicos Silva*, which preceded the *Circo Matemático*, comprised three members, two of whom were then very young (Manuel Silva and Laura Silva). Many a time they fully understood the concepts behind the mathematical tricks without at any moment being aware of any math learning.

Quick tap

Effect

Five cards are placed on the table by the Magician's Helper. A Volunteer chooses one, and communicates his choice to the Helper. The Magician now enters the room. The Helper taps each card once. The Magician discovers which card the Volunteer chose.

Method

The five cards are placed on the table according to the layout of the spots of a five, and one of these cards must be a five. Thus, a simple matching is defined between each spot of this card and each of the five cards. When the Helper touches the five card on the table, he must do so in such a way to indicate the chosen card.

Let's take an example: The participant chose the six of spades. As the Helper goes through the cards, he touches the central spot of the five of clubs, thus indicating the chosen card.

One can adapt this trick to seven cards (instead of five), for example, for a better effect.

Mathematics

Each card on the table corresponds to one and only one spot on the card that serves as a map, that is, we have a bijection between cards and spots.

A game with colors

Effect

The Magician shows a deck of cards to the Volunteer and invites him for a game. The Volunteer should choose a color, red or black — let's suppose he chose black. Then, with the deck facing down on the table, the Volunteer and the Magician alternately draw cards, either holding each card to themselves, in a pile, or discarding them to another pile on the table. The aim is to collect cards of the chosen color (black for Volunteer, red for the Magician) without looking at them. All cards (saved or discarded) are held face down.

When the deck is finished, the Volunteer and the Magician turn over the cards in their piles, revealing that the Volunteer only has black cards and the Magician only has red cards.

Method

The deck has to be prepared, with alternating colors. Before the trick starts, the Magician has to take a quick look at the color of the bottom card: the top card will be of the opposite color. Thus, the Magician must choose carefully who begins taking cards from the deck. In our example, if the bottom card is black, the Magician learns that the top card is red and will say, "I'll play first." It doesn't matter whether the cards are saved or discarded, this only induces an extra element of distraction.

Mathematics

It is quite clear that with this arrangement of cards, one player will only draw red cards and the other black cards.

Klein

Effect

Take four cards as follows:

 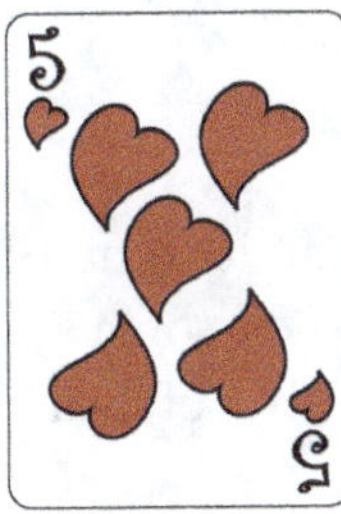  

From those, the Volunteer chooses one and takes it. After memorizing it, he returns it to the four-card deck. The cards are shuffled. The Magician, who was facing away from all this, guesses the card chosen card at a glance.

The Magician can also do this trick without a Helper if she shuffles the cards after the Volunteer returns the card.

Method

The Helper rotates the card chosen by the Volunteer before putting it back in the deck or turns the three-card deck around while the Volunteer looks at the card. Thus, the Magician, who knows the four initial cards, only has to look for one that is oriented differently (note that in the initial position they are all with the center symbols facing upwards). The diamond symbol does not change with a half turn, so if the Magician finds the four cards with the symbols aligned, the Volunteer's card will be the nine of diamonds, otherwise the card with the symbol oriented differently is the card what she is looking for.

Mathematics

Cards are rectangles of paper. The rectangles have many symmetries, that is, there are many geometric transformations that transform a rectangle in itself (for example, mirror reflections):

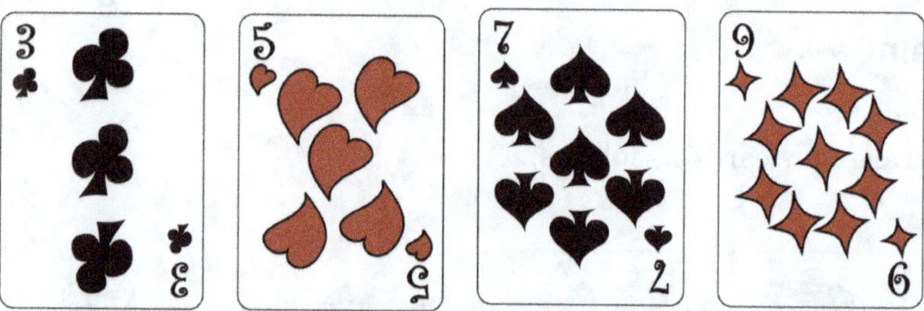

Also, the 180 degree rotation around its center leaves the rectangle unchanged. Nevertheless, the spots spoil some symmetry (except for the diamonds!). Here, we have the five of hearts before and after being rotated:

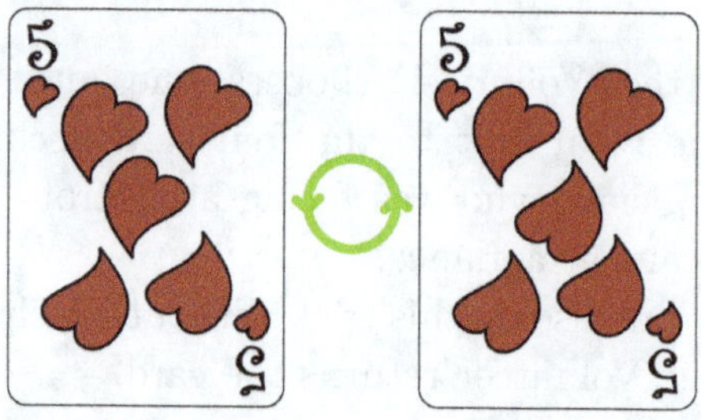

Felix Klein was a German mathematician of the late 19th century and early 20th century who devoted himself to questions of geometry, trying to describe geometric objects through transformations that left them invariant, such as the 180 degree rotation in the case of our rectangle. He was also interested in elementary mathematics, having written a famous collection of books entitled *Elementary Mathematics from a Higher Standpoint*.

One card out of the pocket

Effect

The Magician asks the Volunteer to shuffle the deck and then puts the cards in his pocket — in the end, the Magician will draw a chosen card from this deck. We now present a possible dialogue for this effect.

The Magician asks the Volunteer to select two suits, the Volunteer says hearts and spades. The Magician asks that of these, he selects one; the Volunteer says hearts, and the Magician comments that in this case spades are the suit left.

Next, the Magician asks the Volunteer to select high cards (8 or up) or low cards within the suit of spades. The Volunteer selects the high cards, and the Magician says that in this case the low ones remain. Of these, he asks the Volunteer to select four, the Volunteer says 4, 5, 6, and 7. The Magician asks that of these, she selects two, the Volunteer says 5 and 7. The Magician asks that of these she selects one, the Volunteer says 5, and Magician says that in this case, the seven is left. He then takes the seven of spades from his pocket.

Method

The Magician must take a quick look at the bottom card of the pack before putting it in his pocket and then steer the Volunteer's choices until she reaches that card. In our example, this card was the seven of spades. Thus, when the Volunteer chose hearts and spades, the selection included the right choice, so the Magician took these two choices. Next, the Volunteer chose hearts and the Magician redirected the course to the spades.

Mathematics

The method used in this trick to make the Volunteer arrive at the chosen card is similar to a process that is used in mathematics to find zeros of functions, the bisection method, which consists in dividing successively intervals in half, to approximate a point where a certain function takes the value zero. Later, we will present a somewhat more sophisticated version of this trick, "two out of the pocket".

Four Kings at the inn

Effect

The Magician places a four on the table, representing an inn with four rooms (the four spots), where four kings will be lodged, along with their entourage: a queen, a prince and a page. These characters are represented by K, Q, J, and A. Initially, each group is placed in a room, that is, next to each spot of the four card on the table (see the figure).

Suddenly, in the middle of the night, there is a great commotion, people leave the inn and return to whatever room they can find: The

Magician collects the cards, shuffles them, and puts them back in the rooms, this time facing down. When morning arrives, they find that the groups have been undone, but that there is still some order: The kings are all in the same room, as are the queens, the princes, and the pages.

Method

It is necessary that the order of the cards, in the first distribution, be the same in each room, as shown in the figure. The Magician must collect the piles without mixing them. He then cuts the deck several times — one can do this by holding the deck in hand, executing the so-called cut shuffle (see page 181).

The Magician then turns the deck down and distributes the cards successively to each corner of the four, thus ensuring that all cards of a certain value go to the same stack, that is, to the same room.

Mathematics

The cut shuffle maintains the deck structure: There are always three cards in between a given card and the closest card with the same denomination.

The poison card

Effect

The Magician proposes a poker game to the Volunteer with a 10-card deck. Each player has to keep the hand they receive without any change. The Magician deals and always wins.

Method

The 10-card deck has to be made up of three sets of three of a kind and an extra card — the poison card. The player that gets this card loses. The Magician must shuffle, so that she knows where the card is in order to give it to the Volunteer — she can, for example, leave it over the deck and start dealing to the Volunteer.

Mathematics

One just has to do a case analysis. The hands with the poison card can be a pair, two pairs or three of a kind. The corresponding hands of the opponent are, respectively, two pairs, three of a kind and a full house.

A square of cards

Effect

The Magician shows 16 cards on the table, arranged in four columns with four cards each, face up. She asks the Volunteer to choose a card and point to the column where the card is. She then collects the cards and puts them back on the table, asking again in which column is the Volunteer's card. After this, she presents the chosen card.

Method

After the Volunteer makes his first selection, the Magician collects the columns without mixing the cards and has to remember where she put the chosen column — one way to avoid mistakes is to put it above or below the others. Then, she turns the deck down and places the cards on the table from left to right, in four rows, so that each previous column now becomes a row. If, for example, the Magician has placed the chosen column under the deck (facing up, before turning it over), it becomes the first row. When the Volunteer indicates the column where his card is located, the Magician knows that the chosen card is the first one in that column.

Mathematics

In a square of cards, if we know in which row and which column is a chosen card, we know what the card is.

The Belchou Aces

Effect

The Magician asks the Volunteer to split a deck, facing down, into four roughly equal piles, A, B, C, and D, by means of three cuts. We'll assume the pile D was originally on top in the deck. The Magician then asks him to do the following: Pick up pile A, deal three cards to the place where the pile was and then deal three more cards, one on each pile. The remaining cards are placed in the place where pile A originally was, on top of the three cards that are now there. The process is repeated for the piles B, C and D, in that order.

Now, the Magician turns over the top cards of the four piles, which turn out to be the four Aces.

Method

All four Aces must be at the top of the deck before the trick starts.

Mathematics

Pile D is the one with the four Aces. It receives three cards from the other piles, and then these three cards are dealt to the table, leaving again the Aces at the top.

The book by Gardner (2014) attributes this trick to Steve Belchou and mentions its publication in a 1939 journal.

The four nines

Effect

The Magician invites four Volunteers to participate in the trick. She asks the first one to say a number from 10 to 19, suppose he says 13. Then she draws 13 cards to the table, one at a time. Of these 13, she returns some cards to the deck: First, one card and then the three cards, which are the digits of number 13. She does the same for each Volunteer, leaving four piles on the table. The top cards of these piles are turned, and they are the four nines!

Method

One must place the nines in positions 9, 18, 27 and 36 from the top of the deck, facing down. The Magician must always draw the cards one by one, either to the piles on the table or back to the deck.

Mathematics

The computations that make this trick work are

$$10 - 1 = 11 - 2 = 12 - 3 = \cdots = 19 - 9 = 9$$

As both numbers increase by one unit, the difference remains equal to 9.

Therefore, when taking 13 cards, reversing the order, the first nine becomes the ninth card from below. When you return four cards to the deck, it becomes the top card. In addition, as these four cards return to the deck, the second nine becomes again the ninth card from above, allowing one to repeat the process.

The four eights

Effect

The Magician holds in his hand a deck, facing up, and announces that she is going to make a second pile by dealing some cards from it (one by one). As she deals, she asks the Volunteer to tell her to stop whenever he wants, and at that moment, she stops and places the two piles on the table, facing down.

After asking the Volunteer to choose a pile, she shows the top card from one of the piles, which turns out to be an eight. The Magician puts this eight on the table, facing up, and draws eight cards from the other pile, one by one, without turning them, creating a third pile on the table. Then she reveals the three top cards, which all turn out to be the remaining eights.

Method

Let's consider the deck facing down. The Magician must prepare it by placing two eights at the top of the deck and the other two at positions 8 and 9 from below. When she turns the deck up and draws the cards one by one, she waits for the two eights to pass before asking the Volunteer to tell her to stop. Thus, on the second pile, the two eights are in positions 8 and 9 from above after the pile is turned down.

After the Volunteer chooses a pile, the Magician turns the top card from the first pile, regardless of the choice, and places it on the table, face up. If the Volunteer chooses the other pile, the Magician can say, for instance, "then let's get some cards out of that pile, and to know how many, let's turn the first card of this first one." From this point on, the trick is automatic.

Mathematics

There is practically no mathematics, we only need to note that, when taking the cards one by one from the top of the deck, the order is reversed, and the eights are in positions 8 and 9 from above.

Two Volunteers

Effect

Two Volunteers receive two halves of a deck. They shuffle their share, take a card, memorize it, and trade it with their partner. Each one introduces this new card in its half and shuffles again.

The Magician collects the two halves, joins them, looks at the cards and removes two, which she hides immediately. She asks the Volunteers which cards are chosen and voilà!

Method

Each Volunteer receives a half deck with a special property. The black/red separation would be too obvious. The division between prime numbers $(2, 3, 5, 7, J = 11, K = 13)$ and non-primes $(A = 1, 4, 6, 8, 9, 10, Q = 12)$ yields a nice result (Aces can be divided equally by the two sets to make them both have 26 cards). It is very difficult for a Volunteer to immediately understand the difference between the two halves, but for the Magician, it is very easy to spot the cards that fall out of context.

Mathematics

Among the positive integers $(1, 2, 3, 4, \ldots)$, primes are those that have exactly two divisors, the unit and themselves. For example, 3 is prime (because it is divisible only by 1 and 3), while 6 is composite, since 6 has four divisors: 1, 2, 3, 6. The number 1 because it has only one divisor, it is also not considered a prime.

Two Fibonacci cards

Effect

The Magician gives six cards to the Volunteer, asks him to choose two of them, and tell her the sum of his values ($J = 11$, $Q = 12$, $R = 13$). The Magician guesses the two cards.

Method

The cards used are the following:

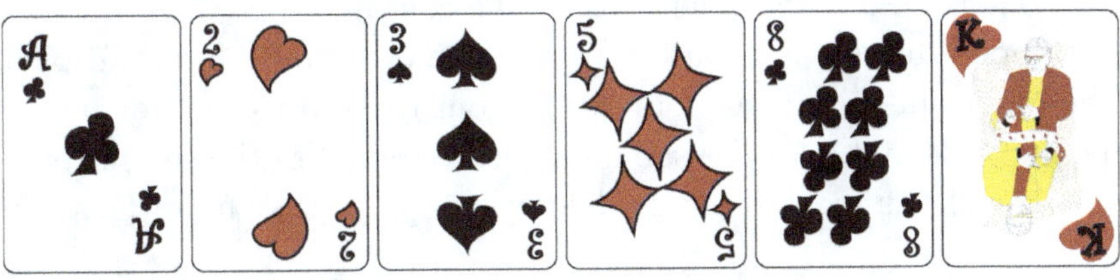

Their values are the first six Fibonacci numbers. Upon learning the sum, the Magician finds the largest number in the sequence that is less than this sum and finds the other by subtracting it from the sum.

For example, suppose the Volunteer announces the sum 11. The largest number in this sequence that is less than 11 is 8 — this is one of the cards. Subtracting 8 from 11, we learn that the other card is 3.

The suits are in order, which is condensed into the well-known word CHaSeD, for clubs, hearts, spades, and diamonds. The word is a mnemonic for the order of the suits.

Mathematics

The values of the cards are Fibonacci numbers: Each number is obtained by adding the two previous ones. This sequence has the following property: The sum of two elements in this set identifies the two elements.

This trick is described in Mulcahy (n.d.).

Lucky piles

Effect

The Magician shows the Volunteer two piles with six cards each: One has the Ace to six of diamonds, and the other, the Ace to six of spades. She places the two piles on the table, facing down.

On the table there are also five coins. The Magician asks the Volunteer to distribute the five coins by the two heaps, as he wishes — he can put all of them on a heap, or four in one and one in the other, or two in one and three in another. This is considered a bet: If the two cards determined by this choice have equal value, the Volunteer loses one coin, otherwise he wins all the coins.

Suppose he sets four on the first pile and one on the second pile. The Magician then takes four cards from the top to the bottom of the first pile and does the same for just one card in the second pile. She then shows the two top cards, revealing that they have the same value.

The Magician removes these two cards from the piles and repeats the process, now with piles of five cards and four coins — the Volunteer just lost one. In the end, the same thing happens again:

The cards from the top of the piles have the same value and are taken away, along with a coin.

The process is repeated until there are only two piles of one card, which will also have the same value.

Method

The trick is automatic, one just needs to order the cards in the initial piles correctly, and be careful when making the cards circulate.

The piles, initially, have to mirror each other: In one of them, the order is from Ace to six, and in the other, it is from six to Ace.

Passing cards from top to bottom of the deck must not reverse the order. The Magician must draw all the cards together, or one by one, and place them under the pile without inverting the order.

Mathematics

Since initially the cards in each pile are mirrored, taking t cards from one pile and $5 - t$ cards from the other leads us to two cards with the same value.

After removing these cards, what happens is that the order remains mirrored. This happens for two reasons. First, the cyclic order is preserved by the way the cards move from top to bottom. Second, after passing the cards indicated by the coins and withdrawing the cards of equal value, the bottom card of one of the piles coincides with the top card of the other (these were the neighboring cards of the cards that were removed).

Four piles

Effect

The Magician asks a Volunteer to choose a card and place it face down on top of a small pile of cards on the table, also face down. On these cards, the Magician places the rest of the deck, face down.

The Magician picks up the deck and deals the cards one by one, turning them over, onto a pile. As she does so, she says the numbers from 10 to 1, in descending order, one for each card. If the number she says coincides with the number of the card she turned, the Magician stops and starts a new pile. If there is never a match, she places a face down card on this pile, starting a new one.

In the end, there are four piles of cards on the table, some covered, some with cards facing up at the top. The Magician sums the spots of the visible cards, obtaining a number n. She then discards n cards of what is left of the deck. After that, the top card of the deck will be the chosen card.

Method

It's a completely self-working trick, one must only make sure the initial pile has seven cards.

Mathematics

As the initial pile contains seven cards, the chosen card is placed in the eighth position from below, and thus in 45th place from the top. Now, each pile will cause 11 cards to be discarded, regardless of which card you have on top. If a pile is capped with a face-down card, it actually has 11 cards. If it is not, as the numbers are decremented, the number showing on the top card is the number of cards left to the 11 cards. These come out at the end.

Here is a possible variant: Instead of asking the Volunteer to choose a card, the Magician may have written the prediction, "I have 8 cards in hand," which turns out to be true after the discard process is finished.

Several effects can be created based on this principle. This one is due to Henry Christ.

Three card guess

Effect

A Volunteer is asked to choose three cards and place them face up on the table. Suppose they were the three of clubs, the 10 of spades, and the five of diamonds.

The Magician says that these three cards will make a prediction, provided you make a small adjustment: We have to discard cards onto each of the three cards until you complete the value of each card to 15. In our example, the Magician picks up the deck, face down, and discards, successively, 12, 5 and 10 cards. For instance, for the three of clubs, you can count "4, 5, 6, ..." until you reach 15. The same can be done for the remaining two cards. If face cards appear, they should have the usual values, $J = 11$, $Q = 12$, and $K = 13$. The discarding should be done, so that you can still see the first three cards appearing below the piles.

Now, the cards are ready to make their prediction. The Magician sums up the values of the cards, in our case 18, and says that these cards are making a prediction about the 18th card, which the Magician has to interpret. In this case, she says that the lowest card is a club card, so the 18th card is also a club card. As for the value, we can start with the other black card, the 10, and add one unit because there is another black card, the three, obtaining 11, which corresponds to the Jack. The Magician then draws 18 cards, confirming that the 18th is the Jack of clubs.

Method

After the initial choice of the Volunteer, the Magician discreetly picks up the deck, facing up, and looks at the fifth card from the top. Alternatively, she can spread the cards on the table, face up, for the Volunteer to choose the three cards, taking the occasion to see which is the fifth card (from above). This will be the card that will appear at the end. The Magician only needs some imagination to come up a story that makes the prediction of the three cards come to be this one.

Mathematics

It is similar to the one in "four piles".

The magical sum

Effect

The Magician sets 25 cards in a 5×5 square and faces away, seeing nothing of what follows. The Helper asks the Volunteer to choose a card in the square. He then turns down all the others in the line and column of the chosen card, leaving the original card face up. The Volunteer chooses four more cards, and each time the Helper repeats the procedure. At the end, there are five cards facing up (chosen by the Volunteer), and the Magician, without looking, tells the sum of these four cards.

Method

The sum is always going to be the same as long as the initial square is well thought out. One possibility is as follows: We present only the values, the suits are indifferent, with the convention $J = 11$, $Q = 12$, $K = 13$:

$$
\begin{array}{ccccc}
9 & 7 & 10 & 8 & J \\
10 & 8 & J & 9 & Q \\
5 & 3 & 6 & 4 & 7 \\
J & 9 & Q & 10 & K \\
7 & 5 & 8 & 6 & 9
\end{array}
$$

With this square, the sum is always 42.

Mathematics

The trick works as long as the square corresponds to an addition table. In the example given, the full table is as follows:

+	3	1	4	2	5
6	9	7	10	8	J
7	10	8	J	9	Q
2	5	3	6	4	7
8	J	9	Q	10	K
4	7	5	8	6	9

We include the numbers hidden in the first table, which get added to give the values in the table.

At the end of the trick, there is one number per line and per column, so that the sum of the numbers chosen is equal to the sum of all the hidden numbers (each appears once and only once in the total sum). For example, suppose the following numbers are left in sight (we leave the extra line and column in view for simplicity):

+	3	1	4	2	5
6			10		
7				9	
2					7
8	J				
4		5			

Then the sum is

$$10 + 9 + 7 + 11 + 5 = (6 + 4) + (7 + 2) + (2 + 5) + (8 + 3) + (4 + 1)$$
$$= (3 + 1 + 4 + 5 + 2) + (6 + 7 + 2 + 8 + 4)$$
$$= 42$$

This trick is due to Martin Gardner.

Out of this world

Effect

The Magician places four Aces on the table, in order: clubs, hearts, spades, and diamonds. She hands the deck, face down, to the Volunteer and asks him to separate, without looking, the black cards from the red cards: If he thinks a card is black, he should place it on the Ace of clubs, if he thinks it is red, it goes over the Ace of hearts. The Volunteer separates the cards without turning any card for the moment.

At one point, the Magician asks the Volunteer to start using the other Aces (spades and diamonds) in the same way as indicators of the color of the cards: If the Volunteer thinks a card is black, he should put it on the Ace of spades, if not, on the Ace of diamonds. The Magician can place a Joker, facing up, to mark the point where the change of Aces should take place.

Alternatively, the Magician can perform this separation of cards, following the instructions of the Volunteer. At the end, there are always four piles, one on each Ace.

When all cards are dealt, the Magician asks the Volunteer to examine the piles on top of the Aces of hearts and spades, verifying that got every card right. She then asks him to do the same for the other two piles, proving that the colors are equally correct — the Volunteer got all the cards right!

Method

The deck, facing down, must be stacked as follows: All the red cards must be above, on top of the black cards. The Joker is placed in the middle, facing up, dividing the sections of different colors.

If there is no Joker, the Magician must count the cards that the Volunteer is dealing and ask him to change the placement of cards when he has already discarded 24 cards.

In the end, while the Volunteer analyzes the middle two piles, the Magician must quickly swap the Aces from one pile to another (or the cards over each Ace).

Mathematics

The trick is completely automatic except for the final swap. The author is Paul Curry, who first presented it in 1942 and it is said that he managed to deceive Winston Churchill. The authors of this book saw a presentation of this trick for the first time in the first *Gathering for Gardner Europe*, in January 2015, performed by John Railing.

Counting the Cards

In this chapter, we present the tricks that involve counting cards, whether in the preparation or in the course of the effect.

The date

Effect

The Magician tells the Volunteer that the Queen of hearts wants to meet him. She looks for the Queen in the deck to make sure that it's there, she finds it, shows it to the Volunteer and puts the card back in the deck. She now asks the Volunteer to set a time for the date with the Queen (from one to 12) and she gives him that number of cards from the pack, for example, if he chooses five, the Magician gives him five cards.

Next, the Magician takes 12 cards from the deck, placed like the hour marks on the dial of a clock, saying, "Let's see if the Queen has arrived on time." She turns the card placed on five, which is indeed the Queen!

Method

The Magician must have 12 cards in one hand when she shows the Volunteer the Queen of hearts. The card is put back in the deck, with the 12 cards on top, so that the Queen in position 13 from the top (with the deck facing down).

The Volunteer chooses an hour and receives from the Magician the cards given by that number (the Magician must remove them, of course, from the top of the deck). After this, she places the cards

like the numbers on the dial of a clock, starting at 12 and going backwards.

Mathematics

This is a variant of "O'Henry", framed as the story of a date. The Queen is initially in position 13. If the chosen time is 5, then the Queen is in position $13 - 5$ from the top after the Volunteer receives five cards. The placing of the cards in descending order puts the card in the right position.

Coins in the pocket

Effect

The Magician places some coins on the table and asks two Volunteers to share them with each other (each one must get at least one coin). Each Volunteer should secretly count how many coins they have and put them in their pocket. This number of coins will be the secret number of each of the Volunteers.

The Magician then picks up a downward-facing deck and announces that he will draw cards one by one from the top and place them face up in a pile on the table, counting aloud as he deals. Each of the Volunteers will have a secret card, which will be the card dealt when their secret number is spoken. For example, if a Volunteer has seven coins, her secret card is the seventh to be turned.

After doing this, the Magician hides the deck and says he will organize the cards in a special way. After shuffling, he presents two piles, facing down, and draws cards from these piles, one by one, simultaneously. From one of the piles, the cards are turned face up and face down from the other. He asks the Volunteers to speak up if any one of them sees their card. When one of them speaks, he stops drawing cards. The last card drawn from the other pile is the card from the other Volunteer.

Method

The Magician initially puts 21 coins on the table. When he shows the cards, so that each Volunteer knows which one is theirs, the Magician can stop when he reaches 20. Then, he turns this small pack down again and places it on the rest of the deck, keeping the cards in the top position.

While hiding the deck, with his left hand, the Magician draws 10 cards, one by one, to the right hand, facing down. You can increase this pile by adding cards from below the left-hand pile, as long as you do not disturb the 11 from the top of the left hand pack.

Now, place the two piles on the table, facing down. From this point, the trick is automatic.

Mathematics

Before beginning the explanation, we note that in a deck with d cards, if a card is in position k from the top it must be in position $d - k - 1$ from the bottom.

Say, one of the Volunteers has n coins in his pocket, the other will have $21 - n$. Since one of the Volunteers must have less than 11 coins, let's assume that $n < 11$. Thus, the two cards chosen are in positions n and $21 - n$ from the top. After dealing the top 10 cards, reversing their order, the card that was in position $21 - n$ is now in position $21 - n - 10 = 11 - n$. The other card, which is now in the small pack, is in position n from below, which corresponds to the position $10 - n + 1 = 11 - n$ from above. That is, both chosen cards chosen are in the same position.

Colorful piles

Effect

The Magician presents a deck to the Volunteer, asks him to shuffle the cards and separates them into two piles of 26 cards, one facing up and one facing down.

She then proposes a game to the Volunteer: She gives him the pile facing up and asks him to separate the black cards from the red ones, one by one, onto two piles, which we will call black and red. The Magician picks up the pile of cards facing down and draws the cards at the same pace as the Volunteer: Every time the Volunteer puts a card on a pile, black or red, the Magician puts her card on a pile in front of it, always facing down. At the end, there is a red pile and a black pile, each with a corresponding pile of cards facing down in front of him.

The Magician then says that the colors of these piles magically influenced the ones that are facing down. How? By influencing the number of cards of each color. Taking the face down piles, and counting, the Magician verifies that the number of black cards in the pile in front of the black pile is equal to the number of red cards in the pile in front of the red pile.

Method

This is a completely self-working trick.

Mathematics

Let b be the number of cards in the black pile and r the number of cards in the red pile. In a complete deck, we have $b + r = 26$ because the Volunteer was left with half a pack. The piles of the Magician (facing down) have $26-b$ black cards and $26-r$ red cards distributed on both sides.

One could now argue in this way: These numbers actually allow for the pile in front of the red pile to have only black cards, in which case the pile in front of the black pile would have only red cards. Then the number of red cards in the pile in front of the red pile and the number of black cards in the pile in front of the black pile would both be zero. Any other disposition of cards in these two downward piles can be achieved by exchanging cards between the two piles, and this operation preserves the equality.

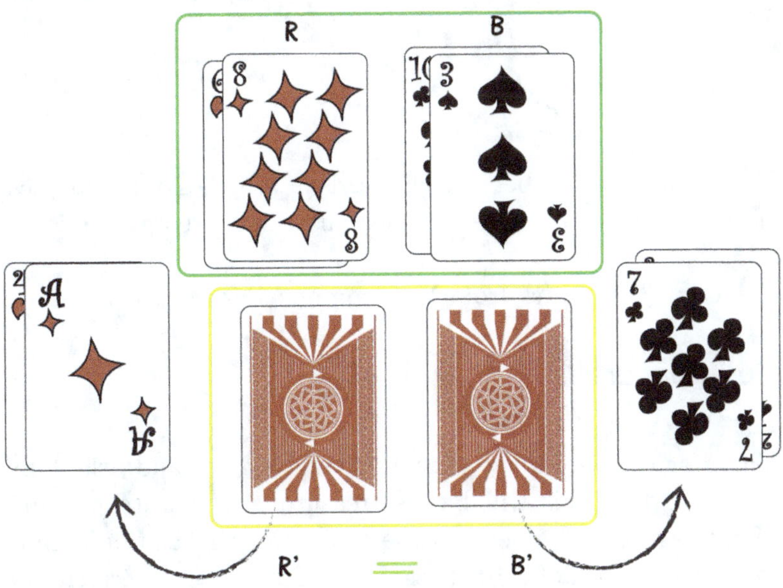

Alternatively, one can also do some algebra. The pile that sits in front of the black pile has b cards of mixed colors. Let b' be the number of black cards in this pile, $b' \leqslant b$. Similarly, the pile in front of the red pile will have r cards, of which r' will be red and $r - r'$ will be black. We want to show that $b' = r'$.

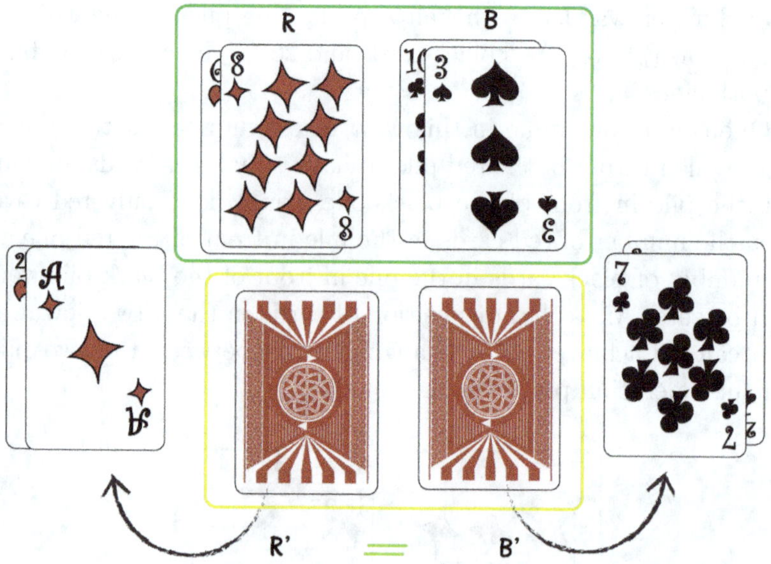

If we count the number of black cards in the piles of the Magician, we have $b' + (r - r') = 26 - b$, which, after some reorganizing, gives us $b' = r'$.

We learned this trick from McOwan and Parker (2010).

Buried treasure

Effect

The Magician gives the deck to the Volunteer and asks him to choose a card — this will be his treasure, which is to be buried. Placing the rest of the deck on the table, the Magician asks him to remove a small pack of cards (less than half the deck), secretly count them and put them in her pocket, remembering the number.

The chosen card will now be placed inside the deck. This number will act as a treasure map, as it indicates to the Volunteer how many cards he must place over her chosen card. For example, if the Volunteer took nine cards, he must place her chosen card under the top nine cards of the deck on the table. The Volunteer does this without the Magician seeing, so that he does not know the secret number.

The Magician then picks up the deck, and says he will try to find the treasure, even without the "map". To do so, she flips several cards from the top of the deck while saying that she is trying to feel, with his thumb, where the chosen card is. At some point, he stops and presents the chosen card.

Method

The Magician has to pass down as many cards as the deck initially had. If it was the full deck, then he must pass 52 cards. It can be done by passing four blocks of 10 cards and two blocks of six, for instance. Note that the blocks are passed without reversing the order of the cards. When passing a block of 10, for example, the Magician must pick up in the deck with his left hand, count 10 cards and take them *in block* in his right hand, not one by one, before adding them to the bottom of the deck.

After this process, the chosen card will be the top card from the deck.

Mathematics

This effect is based on the following simple principle. If we have a deck with n cards, with the positions numbered 1 to n from top to bottom, and we move m cards from top to bottom, then a card that is in position p is now in position $p - m$ modulo n (see page 186). That is, the position of each card decreases m units unless $p < m$, in which case we must add n (the total number of cards in the deck) to get the value of the new position.

If the starting deck has 52 cards and the Volunteer draws k cards, then the final deck has $52 - k$ cards and the chosen card is in position $k + 1$ from the top.

By passing 52 cards from top to bottom, all positions decrease by 52 units, modulo $52 - k$ (which is the number of cards in the pack). Thus, the new position is

$$k + 1 - 52 = 1 + (52 - k) \equiv 1 \quad (\text{mod } 52 - k)$$

This means it is in position 1, the card is on top of the deck.

This effect is described in MacTier (2000).

Spelling bee

Effect

The Magician asks the Volunteer to shuffle the cards and then makes three piles of three cards, face down. She then asks the Volunteer to choose one of the piles and lift it to reveal the bottom card — let's assume it was the eight of diamonds. She then collects the three piles, forming a deck of nine cards, which she holds in his left hand, saying she will call for the card, spelling out its name. She does this by successively passing cards down the deck. In our example, she says, "e-i-g-h-t," and draws five cards, one by one, to the table (or her right hand) and adds them to the deck underneath. She does the same with the word "o-f," drawing and adding two more, and finally saying "d-i-a-m-o-n-d-s," drawing eight more and adding them to the deck underneath. She then turns over the top card, which will not be the chosen card. Simulating surprise, she says that a little magic is missing. She then spells the word "m-a-g-i-c", drawing five more cards to the table: The fifth is the eight of diamonds.

Method

The pile chosen by the Volunteer must be put on top of the other two in the initial pick-up. From this point on, the trick is automatic.

Mathematics

After putting together the pile of nine cards, the first time cards are drawn, at least three cards go to the bottom of the deck (the names of the cards all have three letters or more). The chosen card, which was the third from the top, becomes the third from the bottom. With the word "of", it becomes the fifth card from the bottom, i.e., the middle card of the nine-card deck. The suit names all have five letters or more, so at least five cards are drawn from the top of the deck, keeping the chosen card in the middle position of the deck. Since the word "magic" has five letters, the card that matches the last letter will be the chosen card (any five-letter word can be used).

The general principle is as follows: If you have a chosen card in a deck, and you draw cards to the table one by one until the chosen card comes out, and then put the rest of the deck on top, then the number of cards initially under the chosen card becomes the number of cards on top, and vice versa. If it happens to be the middle card, it will remain in that position. This principle is also used in the trick "three cards and three numbers".

This trick is attributed to Jim Steinmeyer.

Green's "Love"

Effect

The Magician has five cards in his hand. She asks for a number from the Volunteer, let's assume the Volunteer says 9. The Magician passes eight cards down from his deck from the top to the bottom, one by one, turning the ninth card over (without putting it under the deck). She repeats this until there is only one card face up. The name of this card was written on a piece of paper by the Magician beforehand.

Method

You need to place the prediction card on top of the deck. From here on, the trick is automatic. If the number given by the Volunteer has remainder 1 when divided by 5, the first card you turn over is the prediction card (and the effect stops there).

Mathematics

This trick uses five cards, but you could also use any other prime number $(2, 3, 7, 11, 13, \ldots)$. It is named after the Magician Lennart Green, who used to do it with the word "Love".

It is a matter of iterating a sum modulo 5. Since this number is prime, after four iterations of this process, the card facing down is the one that was originally at the top of the deck, as long as the Volunteer's number does not leave a remainder of 1 when divided by 5.

Note that only the remainder of the division by 5 of the Volunteer's number matters, since passing down two cards or seven (or 12, or 17, or...) has the same consequence.

The sum of remainders by 5, called the modular sum (see page 186), has the following table:

+	0	1	2	3	4
0	0	1	2	3	4
1	1	2	3	4	0
2	2	3	4	0	1
3	3	4	0	1	2
4	4	0	1	2	3

We number the initial cards from 0 to 4 (0 will be the top card, 4 the bottom card). In the process described, the number chosen is nine, so we are dealing eight cards at a time, and that is the same as dealing three. If we deal cards 3×3, we are turning over the cards given by the successive additions

$$0 + 3 = 3 \qquad 3 + 3 = 1 \qquad 1 + 3 = 4 \qquad 4 + 3 = 2$$

For the other values from 1 to 4, the result is analogous: These successive additions go through the set $1, 2, 3, 4$ in the first four iterations. The card in position 0 (from the top) remains unturned.

Three cards and three numbers

Effect

Three Volunteers are needed for this trick (if there are only two, the Magician can act as a Volunteer). The Magician asks one of the Volunteers to split the deck into three roughly equal piles. Each Volunteer takes a pile and secretly selects a card from it, which they must memorize and place face down on the table, placing the rest of their pile on top (also face down). Thus, there are now three piles on the table, and the chosen cards are at the bottom of each pile.

Now, the Magician asks each Volunteer to choose a number between 10 and 20, which will be their secret number. To mark their pile, the Magician asks each Volunteer to deal as many cards from their pile to the table as their secret number. The cards should be placed on the table one by one. At the end, if there are any cards left over in the pile, they should be put on top; if the pile has fewer cards than the secret number, it doesn't matter, the count is unfinished.

Now, the Magician switches the piles between the Volunteers and asks them to do the above count again because, although they don't know it, the numbers they have chosen are magical and will produce an amazing effect. Finally, he switches the piles once more and asks the Volunteers to repeat the count for one last time.

He now asks the Volunteers to say out loud what their chosen cards were. As each of them turns over the cards from the top of their pile, the cards turn out to be the chosen ones!

Method

The only thing to watch out for is how to swap the piles. Let's call the Volunteers A, B, and C. The first exchange should be A with C and then A with B. The second exchange should be B with C and then B with A.

Mathematics

The first change has the effect of putting pile A in position C, pile B in position A, and pile C in position B. The second exchange puts the piles back in their initial position. Thus, at the end, each pile returns to the place where it was initially.

If the number chosen by the Volunteer is n, the first move places n cards under the chosen card, moving it to the top half of the pile (the numbers chosen are greater than 10, and each pile will have about 17 cards). In the next move, the chosen card is necessarily one of the cards drawn onto the table, as the number chosen by the Volunteer is large enough for this. By placing the remaining cards on top, the card has now n cards over it. Upon returning to the starting Volunteer, these n cards are removed, leaving the chosen card on top.

If the starting Volunteer's number is greater than the number of cards in their pile, the card is on the top at the end of the first cycle, on the bottom at the end of the second and on top again at the end of the third — the effect is still produced.

This trick was featured on Brushwood (n.d.).

The magic of Manhattan

Effect

The Magician asks the Volunteer to take a small pack of cards from the deck (a little less than half, roughly) and to count the cards (the rest of the deck is discarded). Next, she asks him to add up the digits and memorize the card that is in the position given by that sum from the bottom of the deck (placed face down). For example, if the Volunteer has selected 22 cards, the sum of the digits is 4 and he must memorize the fourth card from the bottom of the deck.

The Magician then takes the deck and spells out the phrase "the magic of Manhattan" drawing one card per letter. The last card will be the chosen card.

Method

It's a completely automatic trick. The Volunteer must draw between 19 and 27 cards for the trick to work.

Mathematics

As we have already mentioned, the difference between a number and the sum of its digits is a multiple of 9, in this case it will always be 18. Thus, in the trick, the card will always be the 19th from the top. In our example, choosing the fourth card corresponds in fact to subtracting 3 from 22.

This trick is mentioned in Gardner (2014), which attributes it to Bill Nord.

Numerology up and down

Effect

The Magician asks two Volunteers to break the deck, each taking about half of it. Both of them split their half in two, placing the four piles face down on the table (two for each Volunteer). Turning to the first Volunteer, the Magician asks him to count the number of cards in one of the piles, which will become their pile, while the Magician takes the other one and shuffles it (the Volunteer's pile must have more than nine cards, if this is not the case, the Magician must ask for another cut).

When the count is done, she asks the Volunteer to use some numerology to select a card from that pile. The Magician asks the Volunteer to add up the digits of the number of cards he just counted (a known numerological method), fan out the cards in front of him and see which card is in that position from left to right (in other words, it's the position from the top of the pile, facing down). After this, she asks the Volunteer to put the piles back together by placing his own on top of the one the Magician has just shuffled.

The process is repeated for the second Volunteer, resulting in two piles. The Magician then turns one of these piles upside down and shuffles them, putting the whole deck back together again, with some cards face up and others face down. She then asks the Volunteers to cut the deck as many times as they want and announces that thanks to the numerological process used, she will be able to find the two chosen cards. After going through the deck, the Magician presents the chosen cards.

Method

When shuffling each of the piles, the Magician must memorize the top card of each one before placing it on the table — these will be the key cards. In each of the Volunteers' piles, the selected card is the 10th from the bottom (facing down). When each Volunteer places their pile on top of the pile that the Magician has shuffled, the selected card becomes the 10th from each key card.

After putting the deck together, with cards up and down, the Magician picks up the deck and looks at the cards facing her. When she finds the key card, she knows that one of the selected cards is the 10th from that one, considering only the cards facing her (there are nine cards in between the key card and the selected card). In case there are less than 10 cards behind the key card, the Magician has to make a cut, placing a pack of cards behind the deck.

After finding this card, she turns the deck upside down and repeats the process.

Mathematics

The explanation is similar to that of the previous effect, bearing in mind that each pile is certain to have fewer than 18 cards. This effect is described in MacTier (2000).

Remembering the future

Effect

Nine cards, with values 1–9, are separated and arranged in a deck face up. The Magician takes this small deck and turns it face down, then cuts it up, so that the position of the cards is not known. As an alternative to cutting, she can spell out the word "future" by passing one card at a time down the deck. Calling a Volunteer, she asks them to choose one of the nine cards in the following way: The Magician puts one card at a time from the small pile onto the remainder of the deck, both face down. The Volunteer can say "stop" at any point, at which moment the Magician stops and puts the small deck aside. She then announces that somehow the large deck knows which card is the top card of the small deck. She then asks the Volunteer to split the large deck into several piles, and to count the number of cards in each pile (there may be more than one Volunteer helping), and then casting out nines from the final result, for example, if there are 23 cards in one of the piles, the result will be 5.

The numbers from the various piles are then added up, and once again the nines are cast out. The number obtained is then the value of the top card of the small deck, which the Magician then reveals.

Method

The only care to take is when breaking the small pack at the beginning. The cards are face down, and their values should be from top to bottom, 1-2-3-4-5-6-7-8-9. You have to take exactly three cards from the bottom and put them on top of the deck, so the deck will be arranged as 7-8-9-1-2-3-4-5-6 from top to bottom, considering the deck face down. Flipping the cards with the word "future" has the same effect. The trick is automatic from this point on.

Mathematics

The result of casting out nines from a number is to calculate the remainder of the division of that number by 9. This operation agrees well with the sum: The remainder of the sum of two numbers corresponds to the sum of the remainders (after reducing this sum).

To explain the trick, note that the large deck, without the small deck, has 43 cards, which gives a remainder 7 when divided by 9 — this is the top card of the small deck. Each time a card from the small deck is added to the large deck, this remainder increases by 1, and so does the top card of the small deck, always maintaining the coincidence.

A possible variant is to do this trick with 18 cards in the small deck (two sets of 1 to 9), always numbered in ascending order from top to bottom, face down, with the 7 on top.

This trick was invented by Stewart James.

Kruskal

Effect

All the cards are placed in a row or somehow in a linear order. The Volunteer chooses a number $1 \leqslant x \leqslant 10$. He then starts counting x, starting from the first card, arriving at another card. He reads the value of this new card (Ace=1, face cards=5) and continues counting in this fashion. At a certain point, you can't go any further. The Magician predicted which was the last card.

This trick can be repeated, it works well.

Alternatively, the Magician can draw cards from the deck face up one by one, asking the Volunteer to do the same type of count, paying attention to the card chosen as the cards are drawn. For example, if the card chosen is a six, the next card chosen will be the sixth card from that one. In this variant it is best to do have some practice rounds before performing the trick.

Method

The Magician does the same as the Volunteer. The last card will be, more often than not, the same, for the various choices of the initial x.

Mathematics

This trick is not foolproof, but its probability of success is high (estimated at over 80%). The reason is that the sequences of the

Volunteer and the Magician tend to converge halfway through the journey. From that point on, they will always coincide.

Kraus

Effect

Remembering that a deck has 52 cards, and that around this number the deck has some magic, the Magician asks the Volunteer for a number between 52 and 60 inclusive, which he must tell her. She then gives him half the cards, asking him to shuffle them so as to pass on his choice to the cards, in some way, while she shuffles the other half.

Asking for the cards back, she puts them together into a single deck, which she shows to the Volunteer, holding it in her hand, face down. She then turns the first card over — its value (each face card is worth 10) tells how many cards she will deal to the table, face down. The last card is then turned face up and again indicates how many cards must now be dealt. For example, if a four is dealt, the Magician draws three cards face down, and the fourth face up. The face up cards should always be in view, they can be dealt to a separate pile, for instance.

When the cards run out, the Magician asks the Volunteer to add up the values of the cards face up: This value is the number initially chosen by the Volunteer!

Method

At the start, the Magician places the following cards, of various suits, underneath the deck (face down) from bottom to top: 9, Ace, 2, 3, 4, 5, 6, 7, 8, 9 and 10 — 9 is thus the bottom card, followed by an Ace, and so on.

The Magician gives the Volunteer the top half of the deck, taking the bottom half for herself. She mentally calculates the difference between 60 and the number chosen by the Volunteer and passes that number of cards from the top to the bottom of the deck (always face down). For example, if the number chosen is 56, the Magician must pass 4 cards. After this, if she wants, she can shuffle the top part of the deck, as long as it doesn't affect the prepared part. When restoring the full deck, the Magician's pile should be placed below (as it was before). The trick is automatic from here on.

Mathematics

Suppose the number of cards passed down was n, which means that the number chosen by the Volunteer was $60 - n$. Since the card before these is a 9, none of these cards will be flipped up, since $0 \leqslant n \leqslant 8$. As you progress through the deck, the sum of the face up cards counts the number of cards that have already come out. As the final prepared series has 11 cards, one of the first 10 must be turned over, and will force the last 9 to be turned over as well. Before this 9 is turned over, the sum of the turned over cards is the number of cards already discarded, which is $52 - n - 1 = 51 - n$. With the nine, it becomes $51 - n + 9 = 60 - n$, which is exactly the number chosen by the Volunteer.

This effect is described in MacTier (2000).

The keystone card discovery

Effect

The Magician shuffles a pack of 20 cards, looks quickly at them and holds the deck face down. She then tells the Volunteer that he, without knowing it, is also a magician. She announces that in this deck of 20 cards he has seen a certain card, say, the Ace of hearts, and that the Volunteer will be able to discover it. Since the Volunteer is a beginner, the Magician whispers that she thinks the Ace of hearts is in the first 10 cards, and asks her for a number up to 10, suppose the Volunteer says 6. The Magician then deals six cards, one by one from the top of the deck and shows the seventh, which turns out not to be the Ace of hearts. The Magician, feigning some disappointment, turns the card face down again, puts the six he drew back on top of the deck, and comments that the Volunteer must not give up, for she is just beginning.

He now suggests that perhaps the Ace of hearts is in the second half of the deck, and asks for a number greater than 10, suppose the Volunteer says 13. As before, the Magician draws 13 cards from the deck, one by one, and shows the next card, which again turns out not to be the Ace of hearts. Commenting that the Volunteer is apparently too distracted, the Magician turns the card back down and places the 13 cards on the table back on the deck.

However, the Magician now says that the Volunteer, seeming not to know about magic, actually does know where the Ace of hearts is. Remembering the numbers chosen before, the Magician asks her to calculate the difference — in this case $13 - 6 = 7$. The Magician then draws 7 cards from the top of the deck, revealing that the next card is the Ace of hearts. She ends up reinforcing the idea that the Volunteer is in fact a magician, without knowing it.

Method

The Magician must know which is the top card of the deck (face down). This is the card that will appear at the end of the process. It is also important that the Magician always draws the cards one by one, reversing the order of the cards.

Mathematics

Using the numbers in our example, after the first move, the Ace of hearts (which was initially on top of the deck) stands at position 6. Then 13 cards are drawn. After drawing 6, the Ace of hearts stands on top of the deck. After that, more cards are drawn: $7 = 13 - 6$, which are put on top of the Ace. For other choices of numbers, the same would happen.

The trick is mentioned in Gardner (2014) and credited to Charles Jordan.

Sum and difference

Effect

The Magician presents a deck which, she says, is programmed to do maths. On the table is a card that will have to intervene at some point in the calculations — suppose it was the Jack of clubs. The Magician asks the Volunteer to choose two numbers from 1 to 10, the magic deck will then guess the values of these cards.

To do this, the Magician asks the Volunteer to tell her the sum of the numbers, and then removes that number of cards from the magic deck, one by one, onto a pile on the table. After that, she returns that pile to the bottom of the deck, after turning it upwards, and places the Jack of clubs on top to help with the calculations.

She then asks what the difference is, and simultaneously draws that number of cards from the top and bottom of the deck (as in the milk shuffle), one card at a time.

She now draws the next two cards, also from above and below, which turn out to have the values of the chosen numbers.

Method

The following cards must be placed in the following positions of the deck, face down (the top card is in position 1):

position	2	4	6	8	10	12	14	16	18	20
card	Ace	2	3	4	5	6	7	8	9	10

That is, you must place a card of value n in position $2n$ (you can insert any card in the remaining positions, as varied as possible to hide the order).

When the Magician picks up the pile of cards he dealt to the table, which is initially face down, she must turn it face up and place it underneath the deck in his hand. Once the Jack of clubs is placed on top, afterwards, the trick is automatic.

Mathematics

The trick is based on the following property, which allows one to find two numbers, knowing their sum and difference. If x is the largest of the two numbers and y is the smallest, we have

$$(x + y) + (x - y) = 2x \qquad (x + y) - (x - y) = 2y$$

The first pile has $x + y$ cards. By drawing another $x - y$ cards from the deck, we draw a total of $2x$, and therefore the last card has value x. We must put in the Jack of clubs, so that this card is not discarded, but stays in the deck.

The pile that was already on the table, with $x + y$ cards in it, is now placed underneath. As we now draw an extra cards $x - y$ cards, the card at the bottom of this pile was in the starting position $2y$ and therefore has value y.

This effect was created by Pedro Freitas.

The fingerprint

Effect

The Magician sets a pack of cards, face down, on the table, and asks the Volunteer to select a card that she will discover through his fingerprints. To do this, before the trick begins, she asks him to put a thumbprint on a Joker, which will be left on the table to serve as a model.

She then asks the Volunteer to take a small pile of cards (about a third, less than half) shuffle it, and pick a card from it. He should memorize it, put his fingerprint on the front and back, then put it back in on top of the deck, face down. After this, she asks him to take another pack of cards from the initial deck (again, about a third) and place it on top of the pack with the chosen card on top.

Now, the Magician takes this new pile of cards and says that she will look at each one of them, looking for a thumbprint that matches the one on the Joker. She then browses the cards of the deck, which she holds in his hand, facing him, and then does the same with the deck facing away from him, stopping at a card which he separates and places on the table, saying that this is the one with the correct fingerprint on both sides. After asking the Volunteer which was his card, she shows the card he has just selected, which is the Volunteer's card.

Method

Before the trick begins, the Magician must memorize the card that is in position 26 from the top, with the deck face down. Suppose it was the Queen of clubs. This will be the key card. At the end of the trick, when the Magician browses the cards facing him, he looks for the Queen of clubs and counts how many cards are behind it. Suppose there are six. The Magician then subtracts this 6 from 25, obtaining 19. After the Magician turns the deck away from him, and chooses the 19th card from the bottom.

Mathematics

The Magician must ensure that the first pack has less than 26 cards and the second pack includes the key card.

The situation after the first cut is as in the picture, with the deck facing downwards. The letter K represents the key card, which the Magician has memorized. The letter C represents the card chosen by the Volunteer:

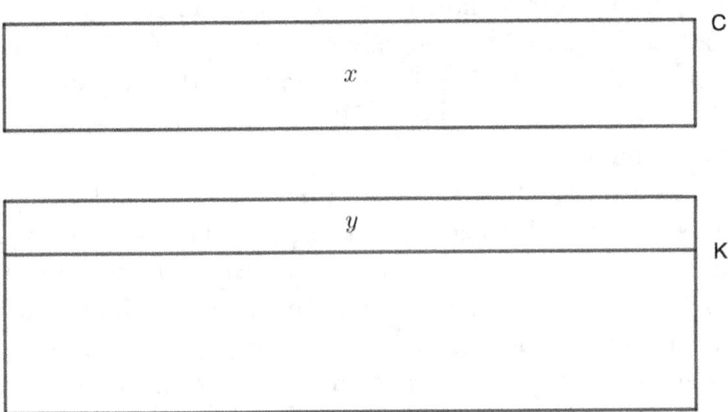

The letters x and y represent the number of cards in the respective packs, excluding K and C. As the key card is in position 26 from the top, we have $x + y = 24$. After adding the second pack on top of the first, the situation is as follows (deck facing downwards):

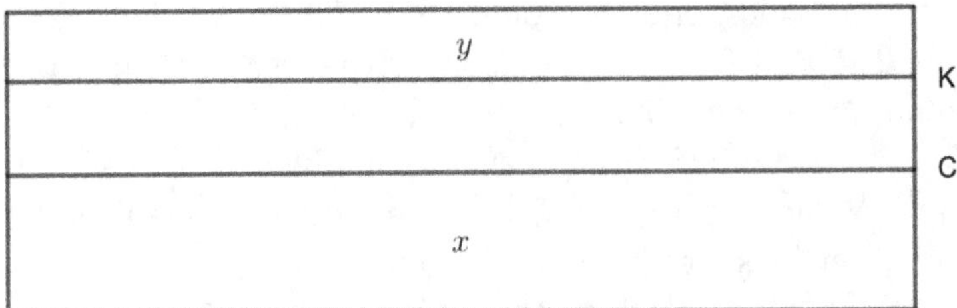

From what we have seen, $x = 24 - y$, which justifies that the chosen card is in position $(24 - y) + 1 = 25 - y$, as the Magician faces the deck.

This trick is an adaptation of an effect found in MacTier (2000).

Two cuts

Effect

The Magician asks the Volunteer to cut the deck at will, twice, forming three roughly equal piles — let's call A the pile corresponding to the top part of the deck, B the middle one and C the bottom one. They should be placed sequentially on the table. In pile B, the top card was determined by the first cut made by the Volunteer — the Magician then asks him to look at it, insert the card randomly into pile B and shuffle it.

Now, we are going to select a card from pile A, but as the Magician may have memorized which one was on top, she asks the Volunteer to shuffle this pile, and only then see which card was left on top, putting it back in the same position. Finally, as the Magician could have memorized the bottom card of the deck, she asks the Volunteer to shuffle pile C. The Magician then gathers them in a deck again. Picking it up, she looks at the cards and draws two cards, which she places on the table, face down. She asks the Volunteer to say which cards were chosen, which turn out to be the two that were selected by the Magician.

Method

The deck must be prepared. Thinking of it face down, the first 11 cards from the top can be any cards, and after those one must place all the spades, ordered from Ace to King, with the Ace on top. So, the 12th card from the top is the Ace of spades, then the two, and so on, up to the King.

After the Volunteer has chosen and shuffled the cards, one must place pile A underneath, pile C in the middle and pile B on top.

When you look at the cards in your hand, you should first go through the cards in block B (these are the cards farthest to the left in your hand) and examine the spades in that block (this is simple as block C has no spades). The lowest spade was the Volunteer's first choice and that is the first one the Magician should put on the table. Suppose it was a 7. To find the second card, the Magician adds 10 to 7, getting 17, and chooses the 17th card from the right.

Mathematics

When the Volunteer makes the cut between piles A and B, it is most likely to occur in the middle of the spades, the piles being roughly equal. So, the card chosen from pile B (which is the top card) is the lowest card of spades in that pile. If the value of this card is n, pile A will have $n-1$ plus the top 11 cards, i.e., $n-1+11 = n+10$ cards. So, your top card is obtained by simply making this sum and counting the cards from the front.

The guessing Jokers

Effect

The Magician sets a pack of cards on the table and asks the first Volunteer to draw approximately one third of the cards and memorize the bottom card of this pile. She then asks the second Volunteer to draw about half of the remaining pile and also memorize the bottom card of his pile.

She then restores the deck, saying that the Jokers, which are the only cards facing up, will help him discover the chosen cards. She then separates the two Jokers from the deck, placing them on the table, and begins removing cards from the deck, one by one, stopping when one of the Jokers "tells" her to stop. Two cards then appear, one next to each Joker, which turn out to be the chosen cards.

Method

The deck must be prepared: with the deck face down there should be 9 cards on top of the first Joker (which is face up) and then 18 more until the second Joker (also face up). From here on the essential thing is how the deck is put together after each cut. Suppose that, in the first two cuts, the first Volunteer removed pile A and the second pile B, leaving pile C on the table. Piles A and B must have one Joker each, which will happen if the three piles are roughly equal. The Magician must place pile A on top of pile C and pile B on top of both piles (always with cards facing down).

The next step is to find the Jokers in the deck and place them on the table, separating in a pile D the cards which were on top of the first and in a pile E those which were between the two, leaving a pile F made up of those under the second. The piles are now grouped by placing F on top of E and D on top of both (thus swapping the positions of piles E and F).

The cards chosen are now 18th from the top and 10th from the bottom (or 43rd from the top). These numbers may possibly be written on papers on the back of the Jokers.

Mathematics

Consider the deck as it is initially prepared: The two Jokers, indicated with J, define three piles, A, B and C.

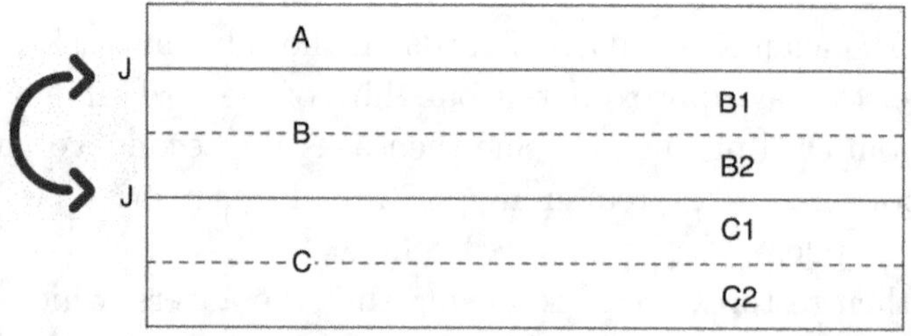

There are 9 cards in pile A, 18 cards in pile B, and 25 cards in pile C. After the Volunteers' cuts (dashed in the picture) piles B and C separate into B1 and B2 and C1 and C2, respectively. The cards chosen are those under B1 and C1. After putting the piles back together according to the rule above (following the arrows), we are left with the following arrangement.

After removing the Jokers and rearranging the deck (again, following the arrows), the situation looks like this.

<table>
<tr><td>B2</td></tr>
<tr><td>B1</td></tr>
<tr><td>C2</td></tr>
<tr><td>C1</td></tr>
<tr><td>A</td></tr>
</table>

As pile B had 18 cards and pile A had 9, the cards chosen are in positions 18 from the top and 10 from the bottom.

To impress the ladies

Effect

The Magician holds the deck in her hand, and says that she will choose a card that will help her — the Jack of clubs. She looks for the Jack in the deck and turns it over, leaving it in the same position in the deck and the rest of the cards face down. Next, she asks the Volunteer to choose a card and memorize it, putting it back in the deck. She now cuts the deck several times, separating it into two piles of 26 cards, saying it will be the Jack of clubs (which is still the only card face up) that will discover the chosen card. To do this, she simultaneously removes cards one by one from the two piles, keeping them face down, until the Jack of clubs appears at the top of one of the piles. When she reaches that point, she turns over the top card from the other pile, revealing it to be the card chosen by the Volunteer.

But there is more: After discarding the Jack of clubs and the chosen card, the Magician says that this effect is always much appreciated by the ladies: Turning over the top cards of the four piles, the Magician reveals they are the four Queens.

And as the Queens do not travel alone, it turns out that the cards under the red Queens are all red, and the cards under the black Queens are all black.

Method

The deck must be prepared initially as follows: Considering the cards facing down, the red cards should be on the bottom, with one

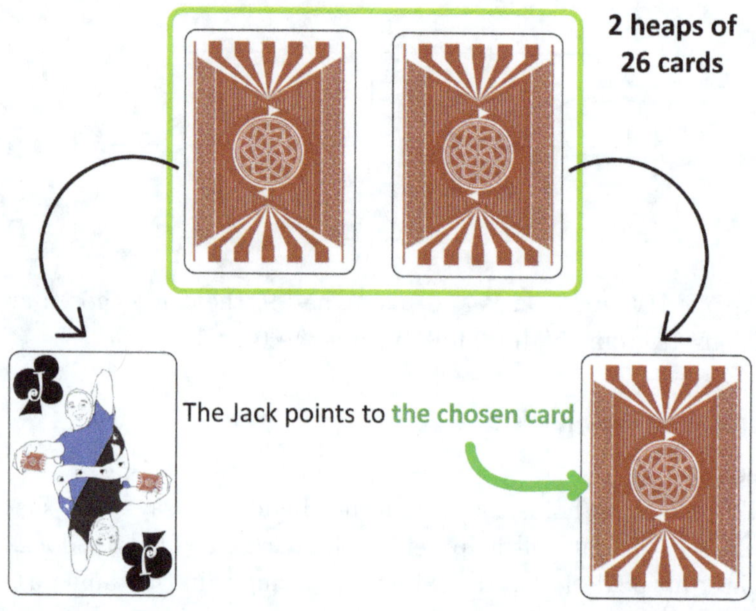

red Queen on top and another underneath. On top of that, one places the Jack of clubs, and then all the black cards, with a Queen on top and another one at the bottom. So, considering the positions of the deck numbered from bottom to top, there are red Queens in positions 1 and 26, black Queens in positions 28 and 52, and the Jack of clubs in position 27. The red cards are in positions 1 to 26 and the black cards in positions 27 to 52.

After turning the Jack face up, the Magician fans only the cards below that Jack, so that the chosen card is red. This card is then placed (face down) on top of the deck. The Magician then cuts several times (he can do the Cut shuffle, for instance) and separates the top 26 cards without inverting the order, that is, cutting 26 cards from the top of the deck and placing them all on the table (and not one by one). From here on the trick is automatic.

Mathematics

The Jack of clubs is 26 cards away from the chosen card, so after the deck is split into two parts, it is paired with the chosen card.

The starting position of the Queens and the red and black cards do the rest.

Five piles

Effect

The Magician gives two piles to two Volunteers and asks each of them to shuffle their pile and place it on the table. Next, she asks each of them to take a pack from their pile, look at the bottom card of that pack, and place it on the table to the right of the rest of the pile.

The Magician then collects the cards and distributes them into five piles. She then takes each of the piles in turn, fanning them to the Volunteers and asking if they see their card, saving the piles that contain the cards. The Magician then examines these piles and presents the chosen cards.

Method

The deck must only have 50 cards (any two can be discarded) and each Volunteer receives 25 cards. The trick is based on the method of collecting the piles and dealing the cards afterwards. Assuming that the two Volunteers stand side by side in front of the Magician, the four piles are in a row, say A, B, C, and D, with the chosen cards placed on the bottom of piles A and C (from the Magician's viewpoint):

Volunteer 1		Volunteer 2	
A	B	C	D

The piles must then be collected from right to left, starting with C: C on top, then B, then A and D at the bottom. This ensures that the card chosen by Volunteer 1 is at the bottom half of the deck and the card of Volunteer 2 is in the top half.

The Magician takes the deck and deals the first 25 cards to five piles successively, one for each pile. When each pile has five cards, she deals five cards to the first pile, five to the second, and so on, until the fifth. This way, the card of Volunteer 1 card is on the top half of the piles whereas the card of Volunteer 2 is at the bottom half.

When the Magician fans the piles, showing the cards to the Volunteers, she memorizes the numbers of the chosen piles. Suppose Volunteer 1 chooses the third pile and Volunteer 2 chooses the fifth. Then the chosen cards are the third from the bottom of the fifth pile and the fifth from the bottom of the third pile. This works even if both pick the same pile. The Magician can make these cards appear in any way she sees fit (for example, she can put the piles in her pocket and take the cards from there).

Mathematics

The initial way of choosing the cards ensures that the distance between the chosen cards is 25, i.e., half a deck. If each half deck is numbered sequentially from 1 to 25 from top to bottom, then the

way of distributing the cards originates the following order in the
five piles:

Pile 1	Pile 2	Pile 3	Pile 4	Pile 5
5	10	15	20	25
4	9	14	19	24
3	8	13	18	23
2	7	12	17	22
1	6	11	16	21
21	22	23	24	25
16	17	18	19	20
11	12	13	14	15
6	7	8	9	10
1	2	3	4	5

The cards directly on the table (in the first position of the first
part of the deck) have their pairs in the first pile. The same goes for
positions 2–5. It is this arrangement that makes the trick work.

This effect is described in MacTier (2000), where it is attributed
to Alex Elmsley.

Penelope's coincidence

Effect

The Magician gives the Volunteer two piles of cards and asks him
to shuffle each one, placing both face down on the table. She then
asks him to choose one of the piles, and see which card is on top of
that pile, memorizing it. After that, she asks him to take some cards
from the second pile and put them on top of the first.

The Magician then takes the first pile and shuffles the cards,
eventually inviting the Volunteer to shuffle the remaining pile as well.
Next, with both piles face down on the table, the Magician invites
the Volunteer to turn over a card from their pile and place it on the
table. The Magician also draws one from her pile, turns it over and
places it on top of the card turned over. Both continue to alternately
draw cards from their pile, placing them face up in a single pile.

The Magician comments that since the Volunteer participated in both choosing the card and building the piles, something magical might happen.

When the Volunteer's cards run out, the card turned over by the Magician will be the chosen card.

Method

The starting decks must have the same number of cards. From this situation, what makes this trick work is the type of shuffling that the Magician uses. After the Volunteer covers the chosen card, the Magician must take the biggest pile (the one with the chosen card, face down). Then she throws a card from the top of the pile to the table (face down) and proceeds to do a milk shuffle (see page 182) on top of that card, always with the cards face down. From that point on, the trick is automatic.

Mathematics

Let us suppose that each pile had 26 cards and that the Volunteer puts an additional n cards on top of the deck with the chosen card. The other deck is then left with $26 - n$ cards.

The Magician then draws the top card, and $n - 1$ further cards paired with the bottom cards, before drawing the chosen card, paired with another card from the bottom. So, at this point, the pile the Magician has in her hand, has one less card on top and n less cards on the bottom, when compared to the initial pile of 26 cards. Therefore, it has $25 - n$ cards, which are now placed on top of the chosen one, leaving the chosen one in position $26 - n$. This is exactly the number of cards as the smaller pile.

The original effect, created by Alex Elmsley, involved a Faro shuffle, which has an identical effect to the perfect shuffle — Elmsley called the original principle the "Penelope Principle". The effect is achieved here by the milk shuffle.

The Elmsley coincidence

Effect

The Magician presents the Volunteer with a pack of cards face down. She asks him to cut it several times, then break it into two piles of roughly equal size and memorize the top cards of these two piles. After that, he can cut and restore each of the piles again, as many times as he wants. With all these cuts, it would seem impossible for the Magician to know anything about the cards seen by the Volunteer, but the Magician says she will still see what she can read off the piles.

She then takes each of the piles, rearranges the cards, and puts them both face down again. She starts to draw cards from both piles at the same time: From one pile, she draws the cards face up from the other pile, face down, asking the Volunteer to stop her when he sees one of the memorized cards. When this happens, the Magician asks the Volunteer for the other chosen card chosen, and turns over the card she took from the other pile, together with this one, which turns out to be the one the Volunteer just mentioned.

Method

Considering the deck face down, the Magician must memorize the bottom card and the one in position 27 from the bottom before the trick begins — these are the two key cards. Since the starting piles produced by the Volunteer are approximately equal, it is very likely that each of these piles will have one of the key cards.

After the cards have been chosen and cuts made, the Magician must count the cards in each of the piles and then cut them as follows:

- If both piles have 26 cards, they must be cut so as to leave the two key cards on top.
- If the piles have different numbers of cards, the smaller pile must be cut, so that the key card is on top. If the difference between 26 and the number of cards in the larger pile is k, the key card of this pile must be in position k from the bottom (i.e., it must have 26 cards on top).

From here on, the trick is automatic.

Mathematics

The distance between the key cards is 26. After the first cut, the piles are like this — let's assume that the second pile was the one that was originally on top of the first.

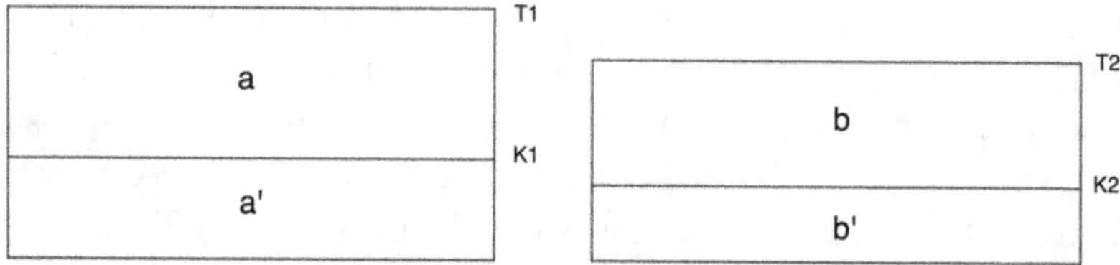

We call the top cards of the first and second piles T1 and T2, which are the chosen ones. The key cards will be denoted by K1 and K2, and are positioned somewhere in the middle of the piles. The letters a, a', b and b' represent the number of cards in each of the blocks, not counting the key cards and the top cards.

Between the key cards, we must have 25 cards, just as we had at the beginning. Thus, $a + b' = a' + b = 24$.

The second pile being the smallest, the Magician cuts this pile in order to put the key card on top. Suppose that the same was done on the first pile. The situation would be as follows:

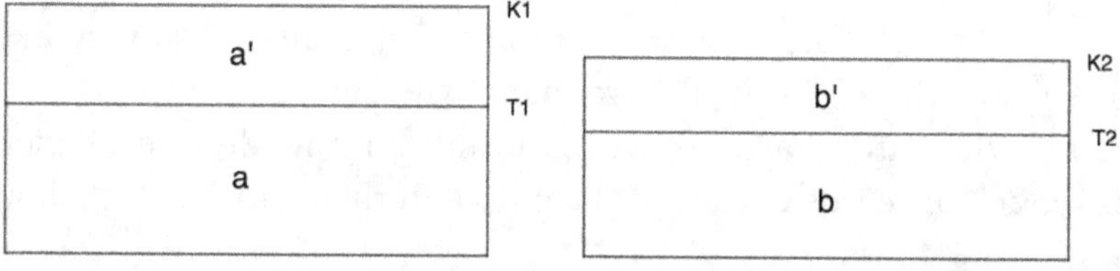

What we need now is to adjust the number of cards above the card T1, so that it equals b'. According to the instructions, we must pass k cards down, where

$$k = (a' + a + 2) - 26 = a' + a - 24$$

and also

$$k = 26 - (b + b' + 2) = 24 - b - b'$$

After this, the number of cards on top of card T1 is

$$a' - k = a' - (24 - b - b') = a' + b - 24 + b' = b'$$

If the number of cards is 26 in both piles, then $k = 0$ and the mathematics still holds.

This effect is mentioned in Gardner (2014), where it is attributed to Alex Elmsley, who published it in a magazine in 1953. This presentation of the effect is due to Dai Vernon.

Shuffle and Deal

In this chapter, the effects are based on different ways to shuffle the cards.

Rainman

Effect

The Magician divides the deck into two parts, placing each pile next to each of her hands. She takes a pack of cards from the right-hand pile, turns it over and shuffles it with the left-hand pile. Now, she takes a pack from the left-hand pile, turns it over and shuffles it with the right-hand pile. After repeating several times, the shuffling ends with a passage of the entire right-hand pile, with a global turn, to the left-hand pile, also with a shuffle.

A Volunteer is called to separate the cards that are face down. These appear listed on a piece of paper that the Magician had written down in advance.

Method

In the end, the cards that are face up are the ones in the left-hand pile, which the Magician knew. This trick, which we learned from Lennart Green, has great impact, although it is very simple.

Mathematics

Whenever a card passes from one pile to another, it is turned over an odd number of times, on the whole, so its orientation is changed. When it goes back to its starting pile it is turned over an even number of times, keeping its orientation.

Australian

Effect

The Volunteer chooses a number between 10 and 30, be it n. The Magician counts n cards from the deck, gives them to the Volunteer to perform an Australian, or down-and-under shuffle (see page 182): One card to the table, one card placed under the deck, repeatedly, until they are down to one card. The Magician guesses that card.

Method

The Magician must know the top card of the deck. Then he must calculate twice the difference between n and the greatest power of 2 less than n. If, for example, $n = 25$, then the highest power of 2 less than 25 is 16, and the number calculated is $2 \times (25 - 16) = 18$.

The Magician then forms a pile with that number of cards (18 in the example), dealing them from the deck, thus reversing their order (first down, second on top of the previous one, and so on). After dealing these 18 cards, the Magician may feign little memory and ask the Volunteer, "How many did you say? Ah, yes, 25," and continue dealing to a different pile (or to his hand) until 25 cards have been drawn. This second pile has to be placed underneath the first pile (if n is a power of 2, this second part is not necessary).

As the Volunteer performs the Australian shuffle, the last card is the one that was on top of the deck at the start.

Mathematics

It is based on a property of the Australian shuffle that we describe in page 182.

Three Australians

Effect

There are three Volunteers. Each one chooses a card and memorizes it: Say, the first chooses card a, the second b and the third c. The Magician deals the remaining 49 cards in the following way: A pile A of 14 cards, plus two, B and C, of 15 each, and is left with five cards in her hand, forming a fourth pile. To achieve this, you can for example take five cards from the deck and then deal cards to three piles: The first two will get 15 cards and the last one, 14.

The first Volunteer should now place card a on the pile A and add some cards from the pile B on it. The second Volunteer should place card b on the (already altered) pile B and then put some cards from the pile C on top of it. The third Volunteer sets the card c on the (modified) pile C, and the Magician puts the five cards from the last pile on the top of them.

The Magician points out that the cards are now lost in the piles because of the cuts that were made at the will of the Volunteers. She then restores the deck, and announces that she is going to try to find them. She takes the deck and does an Australian-style shuffle, for two piles alternately, one face up (the *down* card) and one face down (the *under* card), asking the Volunteers to tell him if they see any of their cards appear. As this does not happen, she takes the face down card pile and repeats the process, asking the Volunteers for their attention. Again, the chosen cards do not appear.

The whole thing is repeated (two more times) until the Magician only has three cards in her hand. Placing them in front of the Volunteers, face down from third to first, she asks each one what their cards were after all. It is revealed that each one has their chosen card in front of them.

Method

The Magician has to collect the piles by placing A underneath, B in the middle and C on top. The rest follows automatically.

Mathematics

Despite the cuts, the cards a, b, and c will always be in the same positions: 38, 22, and 6 from the top. The layout is as follows:

c	
	5 cards
	15 cards
b	
	15 cards
a	
	14 cards

When dealing the cards, the Magician always starts by dealing a face-up card. Thus, the cards in the odd positions are in the face-up pile, and those in the even positions are in the face-down pile. Since the cards a, b, and c are on even positions, they are in the face-down pile, on positions 19, 11, and 3 from the bottom (just divide the previous positions by 2). Since this pile has 26 cards, the positions from the top are 8, 16, and 24 — again, all even. The order is reversed: a above, b in the middle, and c below.

This phenomenon of the cards always being in even positions from the top, will always happen, in all cycles. As there are four rounds in the end the order of the cards is the initial one: c, b, and a from the top.

The warning presented by Diaconis and Graham about the danger of "endless dealing into piles", which we quoted in the introduction, clearly applies to this trick. Thus, in order not to demand too much patience from the Volunteers, this trick has to be used sparingly, and an entertaining conversation should be prepared to accompany the dealing of cards.

If someone wants to use fewer cards, just do the math. If only 26 cards are used, for example, the initial piles must have seven cards (B and C) and two cards (A), leaving the Magician with seven cards in her hand, as can be seen by the positions shown at the end of the first cycle. Which leads to an interesting question: for any given deck, in which positions are the cards left over at the end? The process presented can be applied backwards.

This type of cut that preserves the distance between cards is called *Finnell's Free Cut Principle* in MacTier (2000).

Australian with company

Effect

The Magician presents a small pack of cards to the Volunteer, asking them to choose one and place it underneath the deck, face down. To make this card completely hidden, the Magician takes the deck and draws one more card from top to bottom. She then performs an Australian shuffle, starting by laying one card down on the table, then another down on the deck and so on until all the cards are on the table.

She then takes the deck face down and flips up the top card, stating that this is not the Volunteer's card, but will help her find it. She then deals the cards alternately into two piles. Discarding the pile that does not have the face-up card, stating that she is following the advice of his helper card. She repeats the process twice, until the piles have two cards. Then, the pile that has the two face down cards is discarded, and the other pile kept, which has one card face up and one card face down. This will be the Volunteer's card.

Method

The trick is automatic, provided a 16-card deck is used.

Mathematics

When the number of cards in a deck is a power of 2, the Australian shuffle has the property of placing the two bottom cards of the deck at a distance of half a deck from each other, with the bottom card on top.

Since this deck has 16 cards, the bottom two cards will be 8 cards away after the Australian shuffle — the bottom card is on top and the chosen card is in position 9 from the top. After that they are successively 4, 2 and 1 cards away from each other, that is, in the end it is the two remaining cards in the pile.

Lost and found

Effect

The Magician gives 10 cards to the Volunteer and asks them to secretly select one and place it on top of the deck, face down. She then asks the Volunteer to secretly choose a number up to 10 and pass that number of cards from the top to the bottom of the deck, making this move under the table, away from the eyes of the Magician.

The Magician then takes the cards from the Volunteer, shuffles them, cuts the deck, and returns it to the Volunteer, asking them to deal the same number of cards from top to bottom again. As before, this movement is done under the table. The Magician then asks the Volunteer for the deck and does an Australian shuffle. The last card, which remains in the Magician's hand after the shuffle, is the chosen card.

Method

When the Magician first asks for the deck, she must shuffle it as follows: First, she changes the order of the cards, drawing them from the top one by one, for example, then she moves four cards, in a pack (without reversing the order) from bottom to top.

Mathematics

When the Volunteer passes n cards down the deck, their card is in position n from the bottom. When the Magician changes the order of the cards, the card is shifted to position n from the top. When passing four cards from the bottom to the top, it will be placed in position $n + 4$ if n is less than 7, and in position $n + 4 - 10 = n - 6$ if it is greater than or equal to 7.

In any case, by passing n cards from the top to the bottom of the deck, the card will be in the fourth position from the top. By the properties of the Australian shuffle, the position of the last card remaining after the shuffle is done is given by twice the difference between 10 and 8 (8 is the highest power of 2 less than 10), $2 \times (10 - 8) = 4$.

Soul mate

Effect

The Magician holds a small pack of cards and announces that she can join a card together with its soul mate. As in love affairs, there are always first a few mismatches, the Magician starts by shuffling the cards. She always uses one of the following methods:

(a) Asking the Volunteer for a number from 2 to 4 and deal the cards into 2, 3, or 4 piles, depending on the number chosen. One must deal one card successively for each pile, then collect the piles sequentially (from left to right or right to left).

(b) Making a perfect inverse shuffle, that is, an in-and-out shuffle, separating the cards one by one, without inverting their order, into two piles in your right hand (see page 180). At the end, she puts the pile that is farthest from you on top.

The Magician repeats processes (a) and (b) as many times as the Volunteer asks, interchanging (a) with (b) if this is requested. In order to prepare the deck for the final match, the Magician does a Monge shuffle: She removes the cards, one by one from the pile she holds in her left hand, passing them to her right hand, alternately placing the cards on the top and on the bottom: The second card goes on top of the first, the third underneath, the fourth on top and so on (see page 181 for further explanation).

The deck is now ready to match any card chosen. The Magician starts passing cards from top to bottom of the deck, one by one, until the Volunteer tells her to stop. When this happens, he removes the top card from the deck and places it on the table, face up. Next, the deck will find the pair of that card. To do this, she performs an Australian shuffle and the card left in her hand will be the pair of the card on the table.

Method

This trick is practically automatic, you just need to have the deck conveniently prepared. Six pairs of cards are chosen, for instance, the Ace to 6 of spades and the Ace to 6 of clubs. At the beginning the deck must be mirrored, i.e., with the Ace, 2, 3, 4, 5, and 6 of spades on top, top to bottom, followed by the 6, 5, 4, 3, 2 and Ace of clubs, also top to bottom. This arrangement is called a stay-stack.

The card pairs can be different, of course, as long as the pairing is obvious — some magicians have used cards with the names of famous love pairs, for example.

The Magician can also start by displaying the various pairs, joining them in the deck with the pairs together and perform a Monge shuffle, which sets the deck in stay-stack.

Mathematics

For justifications of some of the following statements, please refer to the section "Ways of shuffling" on page 179.

Processes (a) and (b) preserve the mirror arrangement of the initial deck. For example, if the cards are initially in the order

$$A\ 2\ 3\ 4\ 5\ 6\ 6\ 5\ 4\ 3\ 2\ A$$

by placing them in three piles, they will be arranged as follows:

$$3\ 6\ 4\ A,\ 2\ 5\ 5\ 2,\ A\ 4\ 6\ 3$$

When collecting sequentially, the order becomes one of the following, both in mirror image:

$$A\ 4\ 6\ 3\ 2\ 5\ 5\ 2\ 3\ 6\ 4\ A \text{ or } 3\ 6\ 4\ A\ 2\ 5\ 5\ 2\ A\ 4\ 6\ 3$$

The Monge shuffle puts each card within six cards of its pair. When the top card is removed, the Australian shuffle will leave the sixth card from the top in the hand of the Magician. Now, at that point, the Magician has 11 cards in her hand, the highest power of 2 less than 11 is 8, and so the card left in his hand is the one in position $2 \times (11 - 8) = 6$ as we wished.

In process (a), it is possible to have five piles — you should figure out how to pick up the cards in this case, in order to maintain the mirror arrangement (it cannot be sequential). You can also use a deck of 44 cards because $44/2 = 22 = 2 \times (43 - 32)$, but you should not demand too much of the Volunteer's patience.

This trick is described in Diaconis and Graham (2015), where you can find much more information about the various ways of shuffling.

The whispering Joker

Effect

The Magician shows a small pack of cards and fans it, showing the cards to the Volunteer. She then asks them to place a Joker between any two cards, memorizing those two cards (the trick can also be done with two Volunteers, in which case each one memorizes their card). The Joker must be facing the Magician, i.e., turning away from the other cards.

The Magician then says that she will separate the Joker from its neighboring cards, but that, nevertheless, he will remember these cards. She separates the cards twice, and on the third time, takes the Joker out of the deck, placing it on the table, face up (the rest of the deck is, as always, face down). She begins to remove cards from the deck, two by two, asking the Joker whether or not she should stop. When the Joker "tells" the Magician to stop, she removes two cards from the deck, which turn out to be the ones that were originally next to the Joker. Alternatively, she can put the cards in her pocket and consult the Joker as she takes them out.

Method

This trick is automatic, the secret is in the shuffle and the number of cards in the deck, which must be 20, not counting the Joker.

The Magician always holds the deck face down (the only face up card is the Joker). The first two times he "splits the cards", he deals the cards into two piles, alternately, starting with the left one, then joining the piles with the right one on top.

After doing this twice, he finds the Joker in the deck, places it on the table, and cuts the deck in that position, swapping the pile above the Joker with the one below. Finally, he does a milk shuffle, i.e., he simultaneously draws the top and bottom card of the deck onto the table (for more information on this way of shuffling, see page 182), always asking the Joker if the cards just drawn are the chosen ones. The fifth pair will be the correct one. This process can be done with the cards in the Magician's pocket.

Mathematics

To explain what happens after you separate the cards into two piles, let's assume that the cards are numbered from 1 to 21 from top to bottom in the face-down deck. After separating the cards into two piles, the first pile contains the odd cards and the second pile contains the even cards, in the reversed order. When you put the two piles together (with the second one on top), the order is as described in the second row of the following table:

1	2	3	4	5	6	7	8	9	10
20	18	16	14	12	10	8	4	4	2
3	7	11	15	19	2	6	10	14	18

11	12	13	14	15	16	17	18	19	20	21
21	19	17	15	13	11	9	7	5	3	1
1	5	9	13	17	21	4	8	12	16	20

That is, card 20 is on top, then 18, then 16, and so on. After repeating the process once more, the positions are as shown in the third row. To get this listing, simply repeat the process used to move from the first to the second row: write 3 under 20, 7 under 18, and so on.

If we now look at any position in this third row, and move five positions to the left, we find the next number — we may have to go around the table. Now, initially the Joker and the two memorized cards were side by side, which means that now the Joker is still between them, but now with four cards in between. For example, if the Joker and the chosen cards were initially on positions 9, 10 and 11, they end up in positions 13, 8 and 3, respectively.

This justifies the last part of the trick, where the Joker is removed and the deck is cut at that point.

Looking further at the table, we note that if we read the second row, two initially neighboring cards are now 10 units apart. If we repeated the shuffling process once more, the distance would be 8, and if we repeated it yet again, it would be 4 — just write two more rows of the table to check this. The Magician can take these numbers into account if he wants to repeat the trick, which allows him to shuffle a different number of times — but be careful not to test the Volunteer's patience.

As a variant, you can also use the in-and-out shuffle (see annex), as long as you always put the in-pile on top of the out-pile. The reader can examine the table describing this shuffle and see what happens, it is similar to the one we presented, but not quite the same.

This effect is due to Charles Jordan and is mentioned in Fulves (2012).

Looking for the Joker

Effect

The Magician shows nine cards to the Volunteer, and fans them face down. She says that these are friends of the Joker, who will help him hide among them. She then asks the Volunteer to replace one of the cards with the Joker, placing the removed card face up on the table.

Suppose that this card was a five. The Magician then says that she will place the cards on the table, in two piles, and follow the advice given by the chosen card, turning over the fifth card that appears. He repeats this process successively, always turning over the fifth card, noting that this fifth card always happens to be facing down. The Magician says that this is due to the cunning of the Joker's companions, who hide themselves. Repeat this process until there is only one card face down. At this point, the Magician fans the cards on the table, revealing that they are in order from Ace to 9, with the card facing down in place of the five. The Magician then says that she has finally discovered where the Joker is, turning it face up.

The story of the trick can also be done by saying that the Joker is a fugitive, and the remaining cards are his accomplices. The card drawn is an accomplice who has been captured.

Method

The nine cards, when fanned facing the Magician, must be in the following order:

$$9\ 7\ 5\ 3\ A\ 8\ 6\ 4\ 2$$

Thus, the 9 should be the bottom one and the 2 the top one, with the deck facing up. This order is simple to memorize: First, the odd cards and then the even ones, in descending order. When you split the cards into the two piles, start with the pile on the left. At the end, this pile must be placed on top of the pile on the right. From then on the trick is automatic.

If the Magician wants, she can also do the trick with 6 or 12 cards, as long as the order is this (odd ones in descending order, followed by even ones in descending order too), always putting the left-hand pile on top of the right-hand one.

Mathematics

This explanation is based on a study of permutations of the elements of a set. We number the positions in the deck from top to bottom. Suppose the Ace is on position one, the two on position two and so on. The following table shows the initial position of the cards, and the positions after one and two cycles (the first row simultaneously indicates the positions and the value of the cards in the initial deck):

1	2	3	4	5	6	7	8	9
9	7	5	3	1	8	6	4	2
2	6	1	5	9	4	8	3	7

This can be easily checked by manipulating the cards. To see what would happen after more cycles just write more lines, following the rule used to go from the first line to the second.

Each cycle induces the following permutation on the cards: (153486729). This should be understood as follows: The card in position 1 goes to position 5, the card in position 2 goes to position 9, and so on.

As the cycles are repeated, a card moves through all positions until it returns to its starting position. For example, 3 starts at position 3, then moves on to positions 4, 8, 6, 7, 2, 9, 1, and 5, only then returning to position 3.

As we see, the starting position of the deck is the one that is obtained at the end of a cycle, applied to the ordered deck. So, at the end of eight more cycles, all cards are in the initial position, all having passed through the fifth position (or any other position chosen), all being flipped over when they pass through there. As the Joker starts at position 5, it is the last card to get to this position.

This effect is based on the principle by mathematician Charles S. Peirce, who studied the permutations induced in a deck by certain forms of shuffling and is described in MacTier (2000).

31 cards

Effect

The Magician gives 31 cards to the Volunteer, asks them to choose one and place it on top of the pile, all cards facing down. The Magician places this pile on top of the remaining cards in the deck, and asks the Volunteer to cut the deck twice: First, she is to remove a small pile, and place it next to the deck, then she removes another small pile, and places it on top of the first one. Each pile removed should have between a third and half of the cards in the deck. The rest of the deck is discarded.

The Magician, after making some cuts in the pile thus obtained, says that she will discover the chosen card, separating successively the cards of the pile. At the end, she is left with a card in his hand: This is the chosen card.

Method

All piles are facing down.

When the Volunteer returns the pile of 31 cards, the Magician must memorize the bottom card, which is visible and will be the key card. Then, when she takes the pile created by the two cuts, she cuts it successively until he sees the card, and makes a final cut, so that the card is at the bottom of the pile.

After this, the first separation is made by holding the cards in the hand and separating them into two piles, dealing one for each pile (thus reversing the order of the cards). The Magician then retains the pile where the last card fell, discarding the other.

Now, she takes this pile and does an in-and-out shuffle four times, always discarding the out pile, which is the one that receives the first card.

In the end, the Magician is left with only one card in her hand, which is the chosen card.

Mathematics

In the initial deck, the chosen card is the top card and the key card is the 31st from the top. In the pile formed after the two cuts, after the cut made by the Magician, the bottom card is the key card and the chosen card is the 31st from the bottom. This is because the first cut, in the complete deck, must be made above the key card and the second below it.

Now, consider the deck numbered from bottom to top: The key card is in position 1 and the chosen card is in position 31. When you separate the cards into two piles, the cards in odd positions are in the pile that receives the last card (the key card). This is the pile that should then be retained. We also know that the chosen card is now in position 16 from the top. By shuffling in-and-out, discarding the out pile, the cards in odd positions are discarded, now counted from the top, and the chosen card is now in position 8. By repeating this shuffle three more times, the chosen card moves to positions 4, 2 and 1.

At that point this card will be alone, as we would need to have 32 cards at the start of the in-and-out shuffle for there to be two cards at this point. In turn, this would require at least 61 cards in the starting pile, which cannot happen as the deck only has 52.

This effect is described on the YouTube channel (Brushwood, n.d.).

Daisy's socks

Effect

The Magician calls for several Volunteers (between 2 and 10, although it is good to have at least 4). She asks each one to freely choose a pair of cards from a deck that can be shuffled beforehand. Each pair represents a pair of dirty socks going into the washing machine, so each Volunteer must memorise the cards they have chosen — they are *their* socks. The Magician collects all the pairs of cards, adds some more, saying that they were other pairs of socks to wash. She shuffles, to represent the washing, and arranges the cards on the table in four rows of five cards each, as if they were on the clothesline. The cards are placed face up for the owners to see. Next, she asks the Volunteers, one by one, which (horizontal) rows their socks are in. As each Volunteer calls out these rows, she guesses the pair and returns it to the owner.

Method

The deck must have 20 cards, arranged in pairs — each card is initially next to its pair. The shuffling can be done by passing an even number of cards from top to bottom of the deck, which keeps this order. Next, the pairs of cards are placed in a pattern on the table, so that the letters in the words UNDUE, GOANO, TETRA, RIGID tell the position of the pairs:

$$
\begin{array}{ccccc}
U & N & D & U & E \\
G & O & A & N & O \\
T & E & T & R & A \\
R & I & G & I & D
\end{array}
$$

We now see how this is done. The Magician begins by gathering the pairs and stacking them. Holding the deck face down, she deals the cards alternately into two piles, then places one on top of the other (alternatively, one can do two Monge shuffles). The aim is that each card is placed 10 cards away from its pair.

Next, she deals three cards, one by one from the top of the deck (reversing their order) and moves the group of three down to the bottom of the deck. She then does the same with two more cards, then three more and finally two more, passing a total of 10 cards downwards. Then she places the cards on the table sequentially, vertically from top to bottom, starting from the left-hand column, forming a grid with four rows and five columns as above. The cards automatically adjust to the pattern given by the given words. Thus, if a Volunteer says their socks are in the first and second rows, they must correspond to the letter N, and are therefore in positions 2 and 4, respectively.

Mathematics

The words are chosen, so that each letter pair appears only once. This set has the following characteristic: If we write the letters in a row, in the same order as the cards are laid out on the table, we get

U	G	T	R	N	O	E	I	D	A
1	2	3	4	5	6	7	8	9	10

T	G	U	N	R	I	E	O	A	D
3	2	1	5	4	8	7	6	10	9

The second row shows the location of the pairs. Note that the first 10 letters do not contain any pairs, which is a special feature of this set of words.

The first shuffle puts each card 10 cards away from its pair. The second shuffle, where cards are passed down, turns the 1-2-3-4-5-6-7-8-9-10 order into the 3-2-1-5-4-8-7-6-10-9 order, which matches the desired letter pattern.

Another known word set is formed by the Latin words MUTUS, NOMEN, DEDIT, COCIS. The set we have used is mentioned in Gardner (2014) as having been used by Stewart Judah to create a method that allows the cards to be placed on the table sequentially (as we indicate), although it does not describe how to shuffle. There is also a set of Portuguese words, CINCO, DEDOS, TENTA, PIPAS, orally transmitted to us by Margarida Pinto, who also introduced us to the effect.

Ace, two, and three

The Magician places an Ace, a two and a three on the table from left to right.

Turning her back to the table, she asks a Volunteer to choose one of these cards and turn it face down. After that, she must also turn over the other two cards, after swapping them.

The three cards are now face down, the Magician turns around, picks them up, shuffles them and places them back on the table, face down. She then asks the Volunteer to guess where is the selected card, turning over the card he points to. Now, you can have one of two cases:

- The Magician says, "You got it right; that was your initial choice."
- The Magician says, "Not that one, this one," and turns over the card chosen at the beginning.

In either case, the Magician gets it right!

Method

The Magician mentally numbers the positions on the table as 1, 2, and 3 from right to left (initially, the Ace is on position 3, the two on 2 and the three on 1). She picks up the cards, so that the right card is on top, the middle one in the middle and the left one the bottom. Then she cuts the deck, so that 10 cards pass from top to bottom (or 7, or 13, or 16, or 19...), without reversing their order. She can pass one, two or three at a time, in a block, successively. After that, she places the top card in the middle position, the next one to the right and the last one to the left, always remembering the positions, numbered from right to left 1, 2, 3.

The Volunteer's guess consists of pointing to a card i which is in position j. If $i = j$ (i.e., if the value of the card matches the position where it is), then it is the chosen card. If $i \neq j$, then the chosen card is in position $k \neq i, j$. For example, if the card revealed is the two, and this two is in position 1, then the chosen card is in position 3 because $3 \neq 1, 2$.

For example, if the Ace of diamonds is chosen, you have, before the cards are turned, the following situation:

After the shuffle, the Volunteer makes his guess, pointing not to his chosen card, but to the three of diamonds.

This reveals to the Magician that the three is not in its proper place. Consequently, the two of diamonds is the leftmost card. The

Magician then knows that the card chosen is the Ace and that it is in the rightmost position.

Mathematics

Let's think about the cards Ace, two, and three, placed on the table from left to right (Ace on the left, therefore) and let's see what would happen if the Volunteer did not exchange the positions of the cards.

Initially, the Magician takes the cards, placing the Ace on the bottom, the two in the middle and the three on top. Then she passes down a certain number of cards. The suggested numbers are all of the form $3n + 1$ that is, they have remainder 1 when divided by 3 — in other words, they are congruent with 1 modulo 3 (see the annex on modular arithmetic). This means that shifting 10 or 13 or 16... from the top to the bottom of the deck is the same as dealing only one card, and the effect is that the top card becomes the two, then the Ace and then the three. Placing the cards as indicated, the Ace is on position 1, the Two on position 2 and the three on position 3 (numbered from right to left). So, they all match their position numbers.

If there is an initial exchange, the only card that will end up in the right position is the chosen card. The two remaining cards have their value swapped with the position. The inversion of the order of the positions (from right to left) is just to make the trick more mystifying and the method less transparent.

This trick is an exercise about permutations of a set of three elements. The set of all permutations of a set has interesting mathematical properties. A permutation of a finite set is a rule which matches each element of the set with another element of the set (which can be the same or a different one) in such a way that no two elements have the same image. This set constitutes a group, i.e., given two permutations, if we perform one after the other we still obtain a permutation; for each permutation, there is another one which has exactly the opposite effect, i.e., what one does the other undoes; the identity map, which matches each element to itself, is naturally a permutation.

This effect was created by Bob Hummer.

Luísa's twist

Effect

The Magician places an Ace, a two and a three on the table from left to right: A, 2, 3, and explains that the positions of the cards are 1, 2, and 3, according to their values. She turns her back and asks a Volunteer to mentally choose one of these cards and turn it face down. After that he must also turn the other two cards over, after swapping them. Now, the Volunteer must switch the position of two cards, as many times as he wishes, communicating aloud his procedure, mentioning the positions of the switched cards (the 2 with the 3, the 3 with the 1, and so on). When he is finished, the Magician turns around and points to the chosen card!

Method

One of the cards must be marked (at the back). Let us suppose that it is the three. From the moves communicated by the Volunteer, the Magician follows the path of this card. For example, if the Volunteer says: 1 with the 2, 2 with the 3, 1 with the 3, 2 with the 1, the Magician follows the three from its original position: on the second exchange, it moved to position 2, and on the fourth to position 1.

In the end, the Magician looks for this card in the position where he expects it to be. If it is there, then it has not suffered the initial exchange, so it is itself the chosen card. Otherwise, if it is in a different position, say position 1, then the chosen card is in position 2 (if the three is in position 2 then the chosen card is in position 1).

Mathematics

This trick is also an exercise about the permutations of a set of three elements. The final position of the chosen card is associated with the initial permutation. To find this out, you just need to know the destination of a card. To do this you just have to follow its path and rely on the mark in its back.

Let's look at an example. Assume that the marked card is the three. The Volunteer chooses the Ace and switches two and three, as in the previous effect, then turns over the cards. Note the subtle lack of the Circus flag on the back, corresponding to the three.

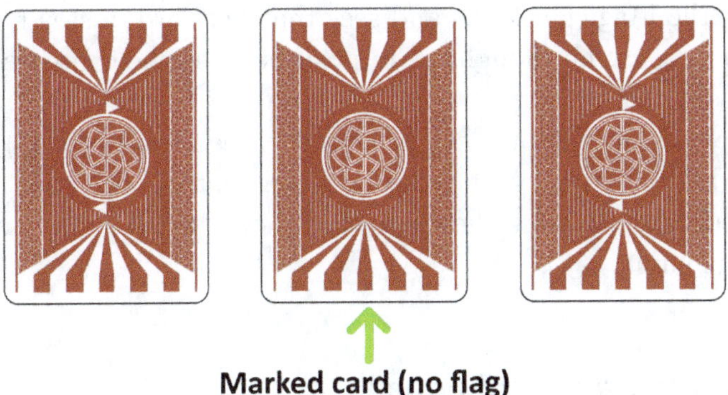

The Volunteer says, "1 with 2" (the Magician thinks, "My card stayed in third position").

Then she says, "1 with the 3" (the Magician thinks, "My card is in the first position") and then, finally, "1 with 2" (the Magician thinks, "My card went to position 2").

Now, the Magician looks and sees that her card, instead of being in position 2, as she thought, is in position 3, so the chosen card will be in position 1.

This is a variant of Ace, two, and three (see page 69), which involves a bit of trickery (it is the only trick in this book that requires a marked card). It was shown to us by our friend Luísa do Vítor, hence the name.

Liberty, fraternity, and equality

Effect

The Magician sets a pack of cards on the table, and tells two Volunteers to divide the cards between them (fraternity), each one taking the cards they want, and shuffle their pile (freedom). The Magician asks each of them how many cards they have, and tells them to give her their cards, saying that in the end she will give them back the same number of cards.

She then turns over one of the piles, so that the cards are facing up, and shuffles them together. After shuffling, she places her hands behind her back, or under the table, with the complete pile and divides the pile in two, giving each Volunteer a pile with the same number of cards that each one had initially. She then asks them both to count how many cards they have face up, finding that they have the same number (equality).

Method

When splitting the starting deck in two, the Magician must memorise the number of cards in the pack facing up. Let's say it has n cards. After shuffling, with the full deck hidden, the Magician separates n cards, which she carefully turns upside down before giving them to the Volunteer who initially had the pack of n cards. The remaining cards form the other pile, which is not turned over.

Mathematics

This trick is the card equivalent of the following well-known trick question. There is a glass of water and a glass of wine and you pass a spoonful of wine into the glass of water, stir and pass a spoonful of this mixture into the glass of wine. Is there more wine in the glass of water or more water in the glass of wine? Since the final volume in each glass is the same as the initial volume, the amounts that went from one glass to the other must also be the same.

When you shuffle the two piles there are n cards facing up. After shuffling, when separating a pile of n cards from the total deck, that pile has m cards facing down and $n - m$ cards facing up. The other pile must have m cards facing up because the total number of face-up cards is n. By flipping over the pile of n pile of cards, both have m cards facing up.

Several effects can be created with this principle. This effect was thought up by Jorge Nuno Silva.

Five/three

Effect

The Magician presents a deck of eight cards, face down, to the Volunteer. She turns the bottom card face up, saying it's meant to hide the value of the cards, and hands the deck to the Volunteer. She then puts two cards on the table, face down, as a prediction.

Then, she asks the Volunteer to perform the following two moves as many times as she wants:

- cut and shuffle the deck,
- turn over the top two cards.

When she stops, the Magician shows the two cards on the table, which are a 5 and a 3. This predicts the arrangement of the cards in the deck: three will face one way and five the other way.

Method

The trick is automatic, provided that, in the deck given to the Volunteer, there are five cards turned to one side and three to the other, or one and seven, as suggest. The trick may eventually fail (with a relatively low probability), in which case the Volunteer will have seven cards turned to one side and one to the other. If the

Magician wants to foresee this case, he may have an Ace and a 7 in her pocket...

Mathematics

After shuffling once, the Volunteer will most likely be left with 3 cards up and 5 cards down. The shuffle does not change the 5/3 pattern (5 up and 3 down or vice versa). By turning the top two cards over, either the situation stays the same (if they are facing opposite sides) or the number of cards facing one side increases by 2, decreasing the number of cards facing the other side by 2. Since $5 - 2 = 3$ and $3 + 2 = 5$, what happens in general is that the 5/3 pattern remains the same.

The trick is due to Bob Hummer and is described in Fulves (2012).

The lady on the train

Effect

The Magician shows a Queen and three number cards to the Volunteer, saying that the three cards represent thieves who are going on a train trip with the lady. Initially, she holds the four cards in her hand, with the Queen underneath, facing downwards, and the three thieves on top, facing upwards (with their backs to the Queen). The lady will try to escape from the thieves during the journey...

Handing the cards to the Volunteer, the Magician asks him to turn over, successively, the top three cards, then the top two cards, and finally the top card, saying "three, two, one" as the train starts to move. Now, the Magician asks the Volunteer to cut the deck (that is, to pass 1, 2 or 3 cards downwards) and then to turn the top two cards upside down. These two movements can be repeated as many times as the Volunteer wants, representing the movements of the characters on the train during the journey.

Finally the train stops (when the Volunteer so wishes). At this point, the Magician asks the Volunteer to make the same movements he did at the start, but in reverse: turn over the top card, then the two top cards, and finally the three top cards, counting "one, two, three." When the Magician asks the Volunteer to examine the pack,

there are three cards facing one side and one card facing the other —
and this card is the lady!

This trick can be done to several people at the same time, each
with their set of 4 cards.

Method

The trick is automatic and requires no preparation; all that is
needed is for the Volunteer to follow the Magician's instructions
scrupulously.

Mathematics

To understand how this effect works, you need to think of the four
cards grouped two by two: The first with the third and the second
with the fourth. The pair of each card will vary throughout the trick,
the companion cards will not always be the same as in the beginning.
It is their position in the deck that tells which card is the companion
of another.

The initial moves have the effect of making the Queen's companion the only card facing the opposite side of the remaining three.

The moves "cut" and "turn the top two" maintain the following
structure in the deck:

- There are always three cards facing one side and one facing the
  other, which we will call the "oddball", and
- The oddball is always the Queen's companion.

To check this, the reader can examine all possible cases, checking
that this structure is always maintained. Essentially, there are three
cases to consider for the "cut" move, depending on how many cards
are passed down, and two cases for the "turn two" move, depending
on whether or not the oddball is among the top two cards.

The last moves turn a card and its companion. Again, an analysis
of the possible cases shows that this turns the Queen into the oddball.

This trick was invented by Charles Hudson, as a variant of a trick
originally created by Bob Hummer. It is described in Diaconis and
Graham (2015), under the name "Baby Hummer". The initial and
final turning movements were suggested to us by Lennart Green, who
also suggested this setting for a trick, involving a train trip.

The Hummer equality

Effect

The Magician gives the Volunteer a pack of cards, asking them to shuffle the cards by cutting the deck and turning the top two cards upside down (this is the Hummer shuffle). The Volunteer can do this as many times as they like until they feel the deck is well shuffled. The Magician then takes the deck, hides it under the table, and says that she is going to try to bring some order to it. She announces after a while that there are as many cards face down as face up, presenting the deck and confirming the prediction.

Method

You need a deck with an even number of cards. When placing the cards under the table, the Magician turns half of the cards over, alternating along the deck: one card is turned, the next one isn't. She can for example pass them successively from the top to the bottom of the deck, turning over every second card, provided he knows how many cards are there in total, in order to know when to stop.

Mathematics

This effect is based on a principle of Bob Hummer, for decks with an even number of cards. As is customary in Hummer's tricks, we will consider the deck divided in two, one formed by the cards in the even positions (from the top) and another by those in the odd positions. The two movements described retain the following structure in the deck: The number of cards facing up, in the even and the odd deck, is the same. To verify this, just consider the cases for the turning movement, and observe that when cutting, either both parts of the deck remain as they are or the even deck becomes the odd deck (and vice-versa).

The principle now is the same as in the effect "Liberty, fraternity, and equality" that we have described. If the total deck has $2n$ cards, the even and odd decks each have n cards. If each of them has m cards facing up, turning one of them face down leaves it with $n - m$ cards face up, which added to the m from the other deck makes up n cards that are exactly half a deck. If the Magician knows how many cards are in the deck, he can also announce how many cards are face up.

The Hummer Sum

Effect

The Magician asks the Volunteer to choose 10 number cards of their choice, and show them to her. Next, after turning over one of the cards, she writes a prediction on a piece of paper and asks the Volunteer to shuffle the cards, cutting and turning the top two as many times as they want (Hummer shuffle). After this, the Magician says she needs to check how well the cards are shuffled, and separates them into two piles, putting them together afterwards. Handing the deck to the Volunteer, she shows the prediction, which says that there are four cards face up and presents the value of their sum. The prediction proves to be correct.

Method

When showing the 10 cards to the audience, the Magician mentally adds up the values of the cards in positions 2, 4, 6 and 8 from the left. Let's suppose that the cards, fanned out, are

$$7\ \mathbf{6}\ 5\ \mathbf{2}\ 8\ \mathbf{7}\ 4\ \mathbf{5}\ 3\ 9$$

The Magician must calculate the sum of the cards marked in bold: $6 + 2 + 7 + 5 = 20$. That will be her prediction. We will call these the selected cards.

Next, she gathers the cards and places them in a small pile face up and turns the top card (in our example, the nine) over. At this point, she passes this deck to the Volunteer for them to do the Hummer shuffle as many times as they like.

At the end, the Magician deals the cards into two piles alternately, observing in which pile the selected cards appear face up (in our example, 6, 2, 7 and 5). When joining the piles, she turns the other pile upside down.

Mathematics

This is another effect based on the Hummer shuffle, "cut and turn two". Consider, as usual, two decks, one made up of cards in the even positions (counting from the top of the deck) and the other of cards that are in the odd positions. Now, consider the four cards chosen. After the initial preparation, what happens is as follows:

- In one of the decks, the chosen cards are face up and the others are face down,
- In the other deck, the opposite is true: the chosen cards are down and the remaining ones are up.

Now, just check that the Hummer shuffle maintains this structure. The final move of dealing the cards alternately to two piles separates the odd and even decks. If the Magician flips over the appropriate pile (the one with the selected cards face down), the cards that are left facing up are exactly the selected cards.

This effect is based on a trick described in MacTier (2000). Note that in the end it is not only the sum that can be predicted, it would even be possible to predict which cards appear. So, if you put the four Aces (for example) in the appropriate positions, they will be the ones that appear face up in the end. The next trick presents such an effect.

The Hummer flush

Effect

The Magician presents the Volunteer with 20 cards, five of each suit, handing him the deck to shuffle. When the Volunteer returns the deck, the Magician asks him which is his favourite suit.

The Magician then takes the pack of cards and begins to set the cards on the table, two by two, in a pile, some upwards, others downwards, saying that her intention is to shuffle them as much as possible. To do so, she tells the Volunteer that he can turn the pack of cards on the table upside down whenever he wants.

When finished, the Magician takes the deck again, as it is, and announces that she will drop small packets of 2 or 4 cards onto the table (forming a pile), in order to shuffle the deck ever further. Before she drops each pack, she asks the Volunteer if he wants it to be dropped as it is, or turned upside down. The Magician follows all the Volunteer's instructions.

At this point, the Magician says she wants to show the Volunteer that the cards are indeed well mixed. To do this, she places them on the table, one by one from left to right, making a rectangle of 4 rows of 5 cards:

X X X X X

X X X X X

X X X X X

X X X X X

Upon reaching the end of each row, the Magician must start again from the left of the next row. As some cards will be face up, others face down, regardless of the suit chosen initially, the Magician takes the opportunity to highlight the confusion.

Now, the Magician invites the Volunteer to help her collect the cards. The method for collecting the cards is as follows: You need to turn each card over, on top of another card that is above, below, left or right. As the process progresses, several small piles are formed, and each pile must be turned over on top of a neighbouring card. This must be done until there is only one pile on the table.

At this point, the Magician asks the Volunteer to recall the chosen suit and spreads the deck, revealing that the cards that are facing up are exactly the five cards of the chosen suit.

Method

The only time the Magician has control over the cards is the first time she throws the cards on the table, immediately after the Volunteer has said what his chosen suit is, and this is when the Magician arranges the deck in a convenient way. The Magician must put down the cards in groups of two, according to the following rule:

- If the two cards are of the chosen suit, they must be set back to back, with the values of both cards showing.

- If none is of the chosen suit, they must face each other (the opposite arrangement).
- If one is of the chosen suit and the other is not, both are turned face up, with the one of the chosen suit on top.

A mnemonic for this rule is "show the chosen suit and hide the others," which is what happens in each case. From here on the trick is automatic.

Mathematics

As with the previous tricks, this one is based on a principle by Bob Hummer. We think of the deck as being made up of two parts: the even positions, and the odd positions (numbered from the top of the deck). The method of setting the cards on the table, which we have described, ensures the following order:

- In the odd deck, the cards of the chosen suit are face up and the others are face down,
- In the even deck, the opposite happens: cards of the chosen suit are face down and the remaining cards are face up.

To convince the reader of this fact, simply place a few pairs of cards on the table, following the rule described. The top card of each pair will go to the odd deck, and the bottom card to the even deck. It is simple to see that the rule arranges the deck in the order described.

It is also simple to see that if you turn any pack that has an even number of cards upside down, the structure is preserved. This is because the cards that are in the odd deck move to the even deck, and vice versa, and also change their orientation. So, even if the Volunteer flips the pack off the table in the first part, or asks to turn over packs of 2 or 4 cards in the second part, it does not change the deck structure.

In the third part, the cards of the odd and even decks are arranged on the table like on a chessboard:

$$
\begin{array}{ccccc}
O & E & O & E & O \\
E & O & E & O & E \\
O & E & O & E & O \\
E & O & E & O & E
\end{array}
$$

At the end of the collection, all the cards are stacked in one position, let's assume it was an odd one. In this case, the cards from the odd deck take an even number of turns to get there, remaining oriented as they were at the start. The cards of the even deck take an odd number of turns, changing their orientation. According to the rule we described, what we got was a deck in which all the cards of the chosen suit are up, and the rest are down.

If by chance the position of the final deck is even, the opposite happens, in which case you must discreetly turn the deck over before spreading it out. In general, as you gather the cards, it is possible to peek and see a card that is facing upwards, to find out whether those of the chosen suit are all up or down.

Here are some possible variants. The Magician can simulate failure by spreading the deck on the table, so that the cards of the chosen suit are down, drawing attention to these only after noting that the cards in view are completely disordered.

You can also use more cards, provided that when you place them on the table, the number of cards in each row is odd, to ensure the final chessboard-like arrangement of the even and odd decks (you have to be careful that the dealing doesn't take too long).

An alternative to the final arrangement is to separate the cards into two piles (one for one side, one for the other, consecutively) and bring all the cards together at the end, turning one of the piles over the other. In this setup, the number of cards in the deck can be any number, as long as it is even.

Instead of five cards of each suit, you can also use a special set of 5 cards in the middle of a 20-card deck, such as a Royal Flush (10, J, Q, K, A of the same suit). This is the suggestion found in Diaconis and Graham (2015).

Secret Codes

In this chapter, we present some tricks based on the ingenious encoding of numbers or messages.

Out of the pocket

Effect

The Magician gives a deck of cards to the Volunteer to be shuffled at his will. Afterwords, the Magician puts it in his pocket, asking him to name a card. Suppose he chooses the 10 of spades.

The Magician announces that he will take a spade from his pocket, followed by two cards that add up to 10, and so he does: First, he takes the four of spades, followed by the two of hearts and the eight of diamonds.

Method

This trick is one of the few that have a little scam. The Volunteer is given an incomplete deck. The Magician keeps in his pocket the following cards: Ace of clubs, two of hearts, four of spades and eight of diamonds, in this order. When the Volunteer returns the deck, the Magician puts the cards in the pocket, so that these four are at the end, or at the beginning, of the deck, to be easily identifiable by manipulation.

If, by chance, the suit cards intervenes in the sum — what happens for example with the six of hearts — then the Magician joins the other cards to the card of the suit, considering the sum total. It may even happen that the requested card is one of four, in which case the effect looks like a miracle.

Mathematics

The suits are all represented in the CHaSeD order, and it is possible to write any number from 1 to 13 using the numbers 1, 2, 4 and 8 — this corresponds to expressing the number as a sum of powers of 2, which is equivalent to writing it in base 2 (see page 185). The effect is due to Findley, and is described in Gardner (2014).

Turning rectangles

Effect

The Magician asks the Volunteer to shuffle the deck. Next, he places some cards, face down, on the table in a rectangular pattern. In our example, we use a square with 5 cards on each side. The Magician turns his back and asks the Volunteer to turn four cards that are at the vertices of a rectangle. The Volunteer can, for example, do the following (we represent a card facing down by × and by 0 a card facing up):

$$
\begin{array}{ccccc}
\times & \times & \times & \times & \times \\
\times & 0 & \times & 0 & \times \\
\times & \times & \times & \times & \times \\
\times & \times & \times & \times & \times \\
\times & 0 & \times & 0 & \times
\end{array}
$$

The Magician tells the Volunteer to repeat the process as many times as he wants — even turning some cards more than once. For example, after turning four more cards, the frame may look like (we show in bold the cards turned on the second turn):

$$
\begin{array}{ccccc}
\times & \times & \times & \times & \times \\
\times & \times & \times & \times & \mathbf{0} \\
\times & \times & \times & \mathbf{0} & \mathbf{0} \\
\times & \times & \times & \times & \times \\
\times & 0 & \times & 0 & \times
\end{array}
$$

After repeating this procedure several times, the Magician asks the Volunteer to choose one of the cards, turn it (if the chosen card is face up it should become face down, if it is face down it becomes face up), and memorize it. Afterwards, the Magician turns himself to the table and guesses the card picked by the Volunteer.

Method

At the end, there is only one row and one column that have an odd number of cards facing up. At the intersection of this line with this column is the chosen card. For example, in the following situation:

$$
\begin{array}{ccccc}
O & \times & \times & O & \times \\
\times & O & \times & \times & O \\
\times & \times & O & \times & \times \\
O & \times & O & O & O \\
\times & O & \times & O & \times
\end{array}
$$

In this case, only the third row and the fourth column have an odd number of facing up cards, so the chosen card must be at the intersection of these lines. The Magician can then simply point it out or collect the cards in such a way that the chosen one is in a known position (for example, above or below the deck), making it appear by any of the usual methods (turning it from the top of the deck, or putting the pack in his pocket and taking it out).

Mathematics

Each time four cards at the vertices of a rectangle are turned, the parity of the cards facing up in each row and column is not changed. This parity was even (zero) at the beginning, so that it remains even until the Volunteer turns his chosen card. This effect is due to Stewart James and is described in MacTier (2000).

Queens and Kings

Effect

The Magician shows eight cards to the Volunteer, the four Queens and the four Kings, asking the Volunteer to memorize one of them. Suppose the Volunteer chooses the Queen of spades. The Magician announces that he will try to discover it without looking at any card. Holding the cards face down, he separates them and shows four to the Volunteer, asking if he sees the chosen one. He repeats this two more times. After this, he makes a pile with 4 cards, another one with 3 and one card stands alone. Turning the card that is at the top of the first pile, the King of spades, the Magician says, "Your card is a spade." Next, he turns the card on top of the three-card pile, which is a Queen, and says, "Your card is a Queen." Then he turns the isolated card which turns out to be the Queen of spades.

Method

The deck must be prepared as follows: First, the four Kings, and then the four Queens, with the same order of suits. For example, a possible order is as follows:

The Magician always shuffles in and out, a card to the top, another to the bottom (see Appendix, page 179, on ways to shuffle). The Magician always shows the *out* pile, the front one, to the Volunteer. In any case, the Magician must put the pile that has the chosen card underneath the others; that is, if the Volunteer says he saw the card he chose, the corresponding pile should go down, otherwise it should go to the top.

After having done this process three times (separate, show and return to collect the pile), the Magician continues with the cards facing down, and draws four cards, one by one from the pile of eight cards to a pile on the table. Then he takes three more, one by one, to another stack, getting one left. The suit and the rank will show automatically correct.

Turning the cards down again, if the piles are collected in order (from left to right or right to left), the deck is again ordered.

Mathematics

Putting always the pack with the chosen card below, this card ends up being the bottom card of the pile, after three iterations (first on the bottom four, then on the bottom two and finally it is the last). Shuffling in and out, done three times, recovers the initial structure: Kings separated from the Queens, and suits in the same order. This makes the fourth card from the top have the same suit as the bottom one, and that the bottom four have the same value (Queen or King).

The trick can also be done with 16 cards (Kings, Queens, Jacks and Aces, for example), calling for four repetitions of the procedure described above. The piles will then have 8, 7 and 1 card, respectively.

Tell me what I know

Effect

The Magician gives nine cards from the Ace to the nine, to the Volunteer and asks him to secretly choose one of them, distributing the rest of the pack by the Magician and the Helper. The Helper does not know which cards were given to the Magician (and vice versa), but he tries to guess which card the Volunteer chose. After that, the Magician confirms the Helper's prediction or corrects it, identifying the chosen card.

Method

The Magician and the Assistant add up the value of the cards they got and calculate the remainder in the division by nine (i.e., they cast out nines). The Magician adds up the amount that the Assistant announces, and the number of the chosen card is the difference between 9 and this value.

For example, if the Assistant says, "I think the card is a 4," and the sum of the Magician is 6, he adds $4 + 6 = 10$, which gives 1 after casting out nines. Then, the Magician knows the chosen card is an $8 = 9 - 1$. If the value obtained by the Assistant or by the Magician is 0, the guess should be 9, of course.

The trick can also be done with an entire suit, that is, with the cards from Ace to the King, but in this case we should use remainders in the division by 13.

Mathematics

We have $1 + 2 + 3 + 4 + 5 + 6 + 7 + 8 + 9 = 45$, which has a remainder of 0 when divided by 9. Thus, if one card is missing, the sum of the remaining cards, after casting out nines, must be a number that added to that of the chosen card is 9, since the total sum is a multiple of 9.

This procedure works with any odd number of cards, n, since in this case $n + 1$ is even and the sum

$$1 + \cdots + n = \frac{n(n+1)}{2} = \frac{n+1}{2} \cdot n$$

which is a multiple of n. So, we consider remainders relative to the division by n.

This effect was created by Jorge Nuno Silva, who adapted it from a problem of the mathematical olympiads.

The programmed deck

Effect

The Magician gives the cards to the Volunteer, asking her to shuffle them. He then says that the deck is programmed to save a certain information that the Volunteer will transmit, as if it were a computer. After looking quickly at the cards, to see if it is "programmed", the Magician returns it to the Volunteer and turns his back.

The Magician then asks the Volunteer to choose a card without saying what it is. Then he says that she will communicate her choice to the deck, coded, passing cards from top to bottom of the deck.

First, he tells her to pass a card from the top to the bottom of the deck if the chosen card is of clubs or hearts. Then, if the card is red, the Volunteer should draw two cards to the table, (one by one) and put the rest on top of them. If the card is black, she should do the same but with three cards.

Finally, the Magician asks that a number of cards equal to value of the chosen card (with the usual convention J $= 11$, Q $= 12$, K $= 13$) should be drawn onto the table, one by one. Once this is done, once again the deck should be put on top of the cards that are on the table.

The Magician then asks the Volunteer for the "programmed" deck, looks at it and guesses the chosen card.

Method

The Magician has to memorize two cards before handing the deck to the Volunteer: the bottom card and the second from the top. These will be the two key cards. It is not necessary for the Magician to remember the starting position of each of the cards, just remember what the cards are.

When the Magician receives the pack, he opens it in the hand, with the cards facing him. He begins by discovering the suit of the chosen card, by looking at the number of cards that are between the key cards. The following table gives the correspondence:

# cards	suit
3	clubs
2	hearts
1	spades
0	diamonds

The order is, as usual, CHaSeD.

Once the suit is known, the Magician counts the cards that are below the second key card. If the suit is clubs or hearts, that's the value of the chosen card. If it is spades or diamonds, it is necessary to subtract one from that number to get the value of the chosen card.

Mathematics

Analyzing what happens with each suit (case by case), we can conclude that the number of cards between the two key cards corresponds to the values in the table above.

In the case of the chosen suit being clubs or hearts, the bottom card of the deck (before passing the cards that encode the rank) is one of the key cards; in the case of of spades or diamonds, there is one more card below the key card. This explains the second rule.

This trick is due originally to Bob Hummer, having been later improved by Ray Grismer, who invented the version that is presented here, taken from Fulves (2012).

The footsteps of the thief

Effect

The Magician gives a Joker to the Volunteer, saying that it is a thief on the run, trying to hide in the middle of a group of indistinct people. The Magician picks up a small pile of cards on top of which the Joker lays. He then asks the Volunteer to cover the fugitive with a number of cards of his choice, by dealing the cards one by one onto the Joker until the Volunteer says "stop". The remaining cards will go underneath the Joker.

The Magician then draws four cards from the rest of the deck and places them side by side on the table, some pushed forward a little. He says that these cards are the thief's footprints and he will use them to capture him.

Then he separates the stack with the Joker in two piles, using the in-and-out shuffle. There are now two separated sets of cards. One of these "falls" on the first "footprint" — the Magician then keeps that stack.

He repeats the process as long as there are "footprints" on the table. After the last footprint, the thief is the card left in the Magician's hand.

Method

This trick is based on the binary representation of numbers, which we describe in the Appendix, see page 185. Any number can be written as a sum of powers of 2, in a unique way. We present below the list of binary expressions for the numbers 0–15.

While doing the trick, the Magician has to count how many cards are above the chosen one, which is done while he draws the cards in front of the Volunteer, until he says "stop". The Magician mentally writes that number in base 2, obtaining the "footprints". Let's look at an example.

Suppose there are 11 cards on the Joker. As $11 = 8 + 2 + 1 = 2^3 + 2^1 + 2^0$, the expression the Magician gets is 1011. Accordingly, the cards are arranged in the following way:

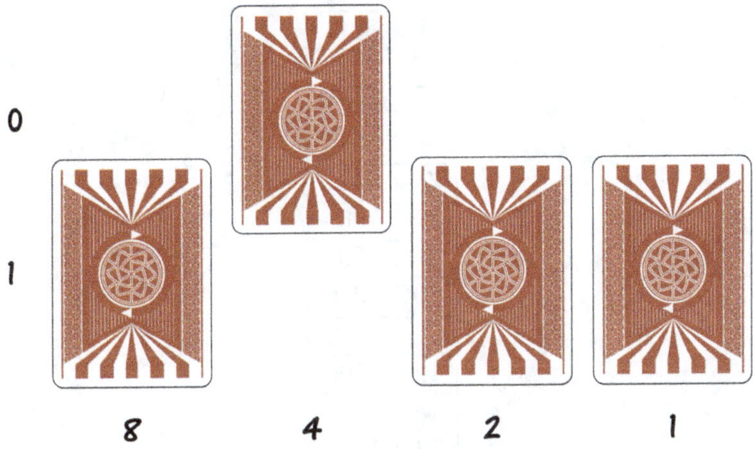

The number 0 is represented by a card above and the number 1 by a card below — this is easy to remember as *in* starts with a letter that resembles 1 and *out* with one that looks like 0.

The Magician should follow the "footsteps" *from the right to the left*. When he separates the cards in different piles, the first card must go to the *out* pile. In our example, *in, in, out, in*.

Mathematics

To explain how the trick works, we have to number the cards from top to bottom. We assign to the first 16 positions the numbers 0 to 15. The chosen card is then in the position given exactly by the number of cards above it (because of position 0). Let's call this position n.

When we make the first separation, the *out* pile gets the cards in which the rightmost digit is 0 — in other words, the even positions. The *in* pile then gets the cards in the odd positions. Thus, for example, if n is even, it will be on the *out* pile, and its rightmost digit will be zero, which will cause the pile chosen to be exactly the *out* pile:

0	0000
1	0001
2	0010
3	0011
4	0100
5	0101
6	0110
7	0111
8	1000
9	1001
10	1010
11	1011
12	1100
13	1101
14	1110
15	1111

If we now remove from the table the rows corresponding to odd numbers (which corresponds to throwing away the *in* pile), we see that the alternating pattern 0/1 is maintained, now in the second to last digit, so that we can apply the same reasoning again. If we remove the lines with even numbers a similar effect happens.

This way of shuffling is also known as *inverse perfect shuffling*.

Erdős

Effect

The Helper gives the following five cards to the Volunteer:

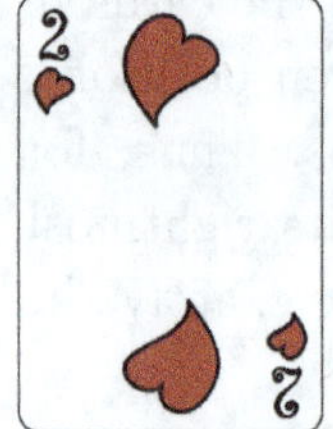   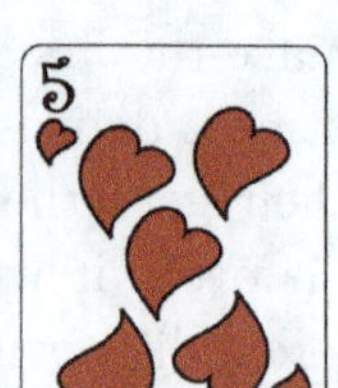 

The Volunteer shuffles them and passes them back to the Helper, who places them on the table, in a row, side by side, face up, stressing that he did not change the order of the cards. The Helper then turns three of of the cards. The Magician, who was away, approaches and guesses each of the cards facing down.

Method

The Magician knows the five cards used from the beginning. The Helper must look at the cards before putting them on the table, and turn three cards in ascending order (left to right). When the Magician arrives he does not have any difficulty in identifying each of the cards, as he knows they are ordered.

Note that the Helper will have to place the cards on the table right to left or left to right, in order to guarantee an increasing sequence of three cards (left to right).

Mathematics

Paul Erdős (1913–1996) was a very famous Hungarian mathematician. Without a permanent address, he traveled relentlessly with light luggage and many ideas to share with his many collaborators. During his lifetime, he published more than 1500 scientific articles, a number that has never been equaled. One theorem of his ensures that given a sequence of five different numbers, three of them (at least) are in order (increasing or decreasing). For example, given $2, 4, 1, 7, 9$, numbers $2, 7$, and 9 appear in increasing order.

In order to prevent that the audience spots a too obvious ordering, Magicians often resort to order alternatives, using for example the order CHaSeD (see page 184).

Let's look at a possible setting for the Helper to leave to the Magician:

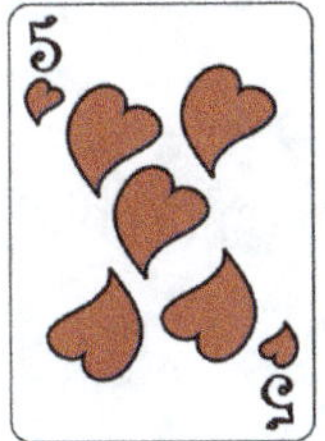

Knowing that the cards covered must be in order (our order!) the deuce will have to be the card to the left, followed by the fives of spades and diamonds, that is,

This trick was presented by *Matemágicos Silva*, who used the following order: deuce of hearts, five of clubs, five of hearts, five of spades, and five of diamonds, see Silva (2008).

Three coins

The Magician lays out the cards from Ace to eight of spades on the table and hands two coins to the Volunteer, asking them to place them on the cards they want (they can even place both on the same card). The Magician then places a third coin on another card and asks the Volunteer to remove any of the coins, remembering the card it was on. The Helper enters the room and guesses where the coin was.

Method

For this purpose, we need to explain the concept of NIM-sum, which we'll denote $\oplus$. To calculate it, simply write each number as a sum of powers of 2, and cut out the repeated powers. For example, $5 \oplus 3 = 6$, since $5 = 4 + 1$, $3 = 2 + 1$ and therefore

$$5 \oplus 3 = (4 + 1) \oplus (2 + 1) = 4 + 2 = 6$$

Alternatively, we can write each number in base 2 (see appendix) and add up the digits, agreeing that

$$0 + 0 = 1 + 1 = 0 \qquad 0 + 1 = 1 + 0 = 1$$

In the example above, 5 is written as 101 and 3 is written as 11 (or 011) in base 2, so

$$101 \oplus 011 = 110$$

and 110 is 6 in base 2.

To get the effect, the Magician adds up the two cards on which the Volunteer placed the coins and places his coin on the card given by the sum, with the exception of 0, in which case the coin should be placed on the 8. For example, if the Volunteer has chosen 3 and 7, the Magician calculates

$$3 \oplus 7 = (2 + 1) \oplus (4 + 2 + 1) = 4$$

and places the coin on card 4.

After the Volunteer removes the coin, the sum of the two remaining cards continues to indicate the card where the coin was. All the Volunteer has to do is add up the marked cards and say the answer.

Mathematics

The NIM-sum has a curious property: In addition to being associative and commutative, like the usual sum, it verifies the property $n \oplus n = 0$, for any number n. So, if $m \oplus n = p$, then also

$$m \oplus p = m \oplus (m \oplus n) = (m \oplus m) \oplus n = n$$

In the same way, we have that $n \oplus p = m$. This property justifies the method used.

Turn one

Effect

The Helper deals the cards Ace, deuce, three, four, five, six, seven, and eight of spades and hearts to the Volunteer. The Volunteer chooses a spade card and informs the Helper of their choice. This card is kept in the Volunteer's pocket and the remaining spades cards are discarded. The hearts cards (from Ace to eight) are laid out by the Volunteer in a row, some with visible faces, others with visible backs, at the Volunteer's discretion, while the Helper announces that, at the end, he will turn over one of the cards, which he does. The Magician enters the room, looks at the row of cards and announces the one in the Volunteer's pocket!

Method

The Helper wants to communicate a number from 1 to 8, which he can do according to the following convention. The number will be given as

$$4A + 2B + C$$

where the coefficients A, B, and C are 0 or 1 (the values $A = B = C = 0$ correspond to the eight of spades, the other correspondences are natural). Let's associate the eight values of the positions of the hearts cards with three parameters a, b, c, as follows:

7	6	5	4	3	2	1	0
a, b, c	a, b	a, c	a	b, c	b	c	
111	110	101	100	011	010	001	000

The third line corresponds to the expression of the number of each position in base two, which we'll explain in an appendix. We take A as the sum, modulo 2, of the number of a's corresponding to face-up cards, and similarly for B and C. By "modulo 2", we mean that the sum is 0 or 1 depending on whether the usual count is even or odd. To calculate the number of a's, b's, and c's, we can see that these letters appear in patterns that are easy to memorise: the letter a appears in the first four positions, the letter b in positions 7, 6, 3 and 2, and the letter c alternately.

Another way of doing the maths is to use the NIM-sum described in the previous trick. The relationship between this operation and the previous method is as follows: The number described by $4A + 2B + C$ is exactly the NIM-sum of the positions of the face-up cards. In fact, the parity of the total number of a's is exactly the parity of 1's in the position relative to 4 because the positions where the a appears are those where the expression in base 2 has a 1 in the position of 4. The situation is similar for the other letters: the b's for the position of 2 and the c's for the position of 1. Just take a look at the table to check this.

Assume the Volunteer chose the five of spades and left the following configuration:

Accordingly to the table we must account for the occurrences of a, b, c corresponding to the positions 7,5,3,2,0. We get:

$$a, b, c + a, c + b, c + b + 0$$

Therefore, $A = 0, B = 1, C = 1$, producing the number

$$4 \times 0 + 2 \times 1 + 1 = 3$$

To obtain 5 we must change the values of A and B, which can be accomplished by changing the card in position 6 (which corresponds to the set $\{a, b\}$):

This configuration corresponds to the values:

$$a, b, c + a, b + a, c + b, c + b$$

so $A = 1, B = 0, C = 1$ and $4 \times 1 + 1 = 5$.

We could also have found the NIM-sum of 3 e 5:

$$3 \oplus 5 = (2 + 1) \oplus (4 + 1) = 2 + 4 = 6$$

We get the same answer: it's the card in position 6 that needs to be changed.

If the cards are placed in two rows, coding and reading can become simpler, as the first digit is given by the rows, the second by blocks of 4 cards and the third by the parity of the columns. Let's give an example.

7	6	5	4
111	110	101	100
3	2	1	0
011	010	001	000

Mathematics

Given a set of three elements, $X = \{a, b, c\}$, we know that it has eight subsets: $\{a, b, c\}$, $\{a, b\}$, $\{a, c\}$, $\{b, c\}$, $\{a\}$, $\{b\}$, $\{c\}$, $\{\}$. The structure of these subsets is such that, for any choice of some of them, it is always possible to add another or subtract one of them, so that the total occurrences of the elements have the desired parity. And the method is simple: given a collection of some of them, if we want to change the parity of the occurrence of some letters a, b or c (to change A, B or C), we just have to choose the subset that contains exactly those letters. If it's in the initial collection, we remove it, if not, we add it. If we're satisfied with the initial arrangement, we add or remove the empty set — $\{\}$ — which is also a subset of X (this corresponds to turning the card over from position 0).

Now, interpreting the situation using the NIM-sum, we can associate each of these subsets with a number, corresponding to its position, as described in the table. To a collection of subsets we match the NIM-sum of their positions, remembering that the number described by $4A + 2B + C$ is exactly the NIM-sum of the positions of the face-up cards. Furthermore, it's easy to see that turning over a card corresponds to matching the NIM-sum of its position with the number described by the row of cards, and this is true whether you turn the card face up or face down.

Let's say we're given the number m and we want to transmit the number n. Since the NIM-sum is associative and verifies $m \oplus m = 0$, we have

$$m \oplus (m \oplus n) = (m \oplus m) \oplus n = n$$

So, the number we have to add to m to get n is $m \oplus n$, i.e., this is the position of the card to be turned over.

Pacioli's two rows

Effect

The Magician places two rows of four cards each, face up, on the table and asks the Volunteer to secretly choose a card and memorize it.

The Magician asks which horizontal row the chosen card is in, collects the cards and puts them back on the table. He repeats this process one more time. At the end, he joins the cards in his hand, making an Australian shuffle. The card that is left in the hand, at the end, is the card chosen by the Volunteer.

Method

There are two ways to collect the cards, always from right to left: you can start with the first row or the second. Starting with the first row, the order is as follows-:

7	5	3	1
8	6	4	2

Starting with the second row, this is the order:

8	6	4	2
7	5	3	1

The Magician collects the cards one by one, placing each new card under the pile that is forming in his hand (it's a natural movement, by the way).

In any case, after collecting, he turns the deck face down before putting the cards back on the table. Starting with the first row of four cards (one at a time, turning them upwards from left to right), then doing the same with the second row.

To collect, the Magician must always start with the row that does not contain the Volunteer's card. When he collects the cards for the third time, the chosen card will be the last to be collected.

Let's look at an example. Here, the chosen card is the two of clubs.

As we mentioned, the cards are collected starting in the row where the two of clubs is not present. After the first collection, the Ace of clubs is the card that sits under the pile, and when it is turned upside down, it becomes the top card.

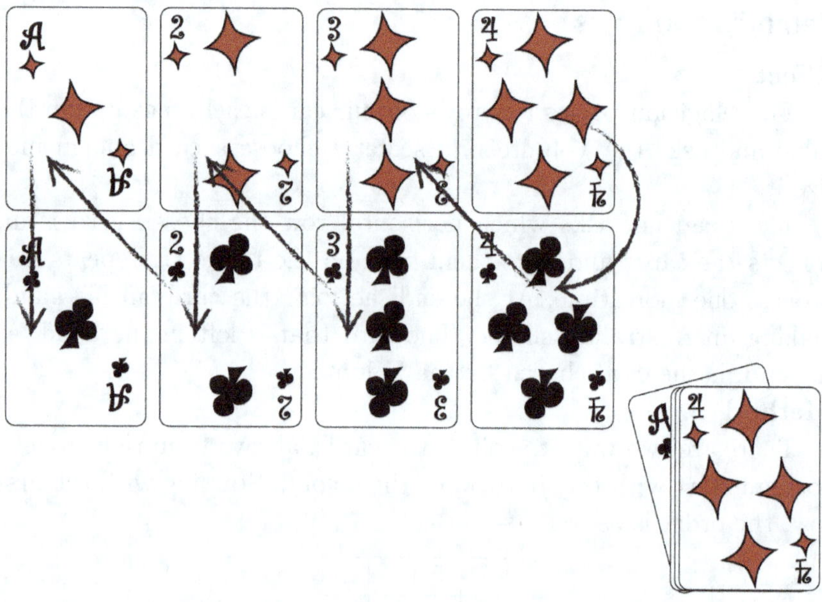

The process is repeated, following the same rule: Start with the line that does not include the Two of clubs.

The process is repeated one last time, with the two of clubs at the beginning of one of the lines.

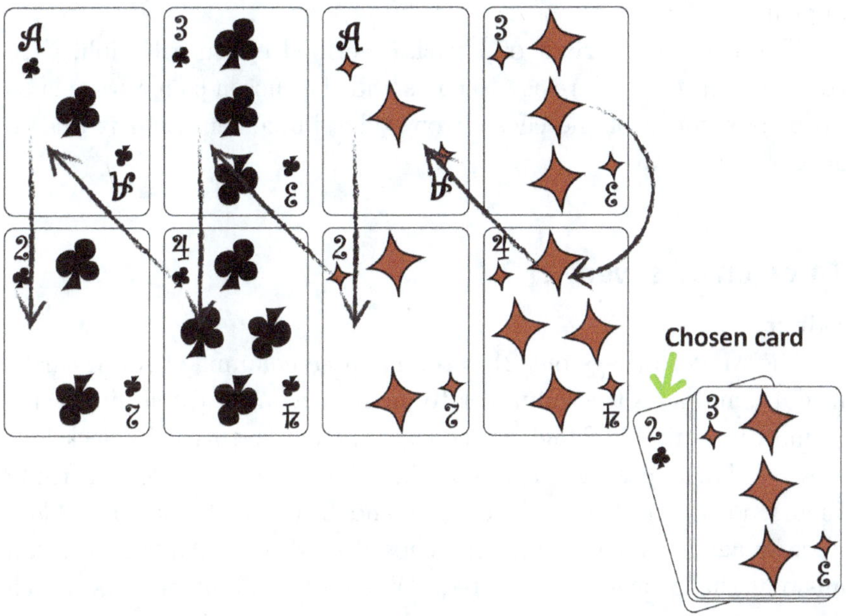

After these three deals and collections, the two of clubs stands at the top of the pile.

When doing the Australian shuffle, the Magician must start by passing a card down from the top of the deck, then deal a card to the table, face up, and so on.

Mathematics

After the first collection, the Magician is left with a deck of 8 cards in his hand, all face up. Because he started with the row that doesn't have the chosen card, this is certainly in an even position from the top. When the deck is turned down, before rearranging the cards on the table, the card is in an odd position from the top. This means that when the Magician lays out the cards on the table again, the card is in the first column or the third column.

After the second collection, the chosen card stands in a position multiple of four from the top before turning the deck over. When the cards go back on the table, the card will be in the first column. After the third draw, the card will be underneath the deck facing up.

The properties of the Australian shuffle then make this card appear.

This trick is a variant of a trick described by Luca Pacioli, with coins. In this case, there are 16 coins and the chosen coin always ends up in position 6 of the chosen row. This implementation is due to Jorge Nuno Silva.

Three times seven, 21

Effect

The Magician lays out 21 cards in three columns of seven cards, face up, and asks the Volunteer to memorize one and tell him which column it is in. The Magician collects the cards, turns the deck face down and arranges them again on the table in three columns, turning them one by one from left to right and from top to bottom. Then he asks again which column the chosen card is in. After getting an answer, the process is repeated. When the Volunteer says which column the card is in for the third time, the Magician says which card it is, or else collects the cards again and announces that it is in position 11 of the deck.

Method

When collecting the cards, the Magician always places the chosen column in the middle of the other two. When he lays out the cards on the table for the third time, the chosen card is the middle card of the chosen column. To make the card appear in position 11, simply collect it by placing that column again in the middle of the other two.

Mathematics

To explain how the chosen card arrives at the middle position, let's look at what happens to the cards each time you do the process of collecting and re-dealing the cards. After the Volunteer points to the column where the chosen card is, this column is placed between the other two. When the cards are arranged on the table, the chosen column ends up in the positions marked with O:

X X X
X X X
X O O
O O O
O O X
X X X
X X X

To check this, simply experiment (with red cards in the chosen column and black in the others, for example).

So, at this point, the chosen card is in one of the middle three rows.

When the Volunteer points again to the column where the chosen card is, this column is placed in the middle of the other two, and as the cards are dealt on the table, the three central cards of this column are in the central positions of the columns (again, the reader can experiment by marking the three central cards of a column).

Thus, this method is an iterative process that places the chosen card in the center of the columns after three iterations.

Three times nine, 27

This effect is an improvement on the previous trick, three times seven, 21. This time, the Magician asks the Volunteer to tell him a number between 1 and 27. He then arranges 27 cards into three columns of nine cards and asks the Volunteer to memorize one and tell him which column it is in. The Magician collects the cards, turns the deck face down, and arranges them again on the table in three columns, turning them over one by one from left to right. He then asks again in which column the chosen card is. After the answer, he does the same for a third time. After collecting the cards, the chosen card is in the position requested by the Volunteer, counting the cards in the deck from the top, with the deck facing down. Alternatively, instead of asking for a number, the Magician can ask the Volunteer to do the whole process by himself, announcing at the end in which position the card is, or by guessing the card itself, which can be done by dealing the cards from the deck one by one and stopping at the chosen card.

Method

The mechanism of this effect is somewhat delicate, but we should not be surprised that it is possible. In fact, each time the Magician deals the cards, the Volunteer tells him which set of nine cards it is in (there are three columns of nine cards at a time). In total there are 27 cards ($27 = 3 \times 3 \times 3$). Imagine the 27 cards divided into three sets of nine, and the Volunteer saying which one his card is in. Then we would take only these nine cards, divide them into three sets of three, and repeat the question. Finally, we had three cards to ask the same question. What happens in this trick is similar, but the mechanism is very well hidden.

We describe two methods presented in the book Gardner (2014). For the first one, it is enough to memorize the following table, which refers to the deck facing up:

	1st cycle	2nd cycle	3rd cycle
Top	3	6	18
Half	2	3	9
Low	1	0	0

This table indicates the number that must be added when the chosen column is placed in the indicated position on the deck (face up). This is a quick way to work with the decomposition in base 3. In the first column, the numbers are 1, 2, and 3 instead of 0, 1, and 2 because it already takes into account the unit that has to be added to adjust the numbering from 0 to 26 to 1 to 27.

For the first effect, we need to decompose the chosen number as a sum of three numbers, one taken from each column (the decomposition is unique: if we do all possible sums, we find all the numbers from 1 to 27, without repetition). For example, if the position is 23, we write, so the column with the chosen card must be placed between the other two, in the first two cycles, and on top of the deck in the last cycle. A practical way to do this decomposition is to start with the summand in the third column, then find the one in the second and finally the one in the first.

For the second effect, simply add the numbers given by each column as the Volunteer arranges the cards, noting what position the Volunteer puts the columns in the deck.

In the mentioned book there is also a method for the first effect that does not need any table, nor any decomposition. This method is due to Thomas Walker. Let's exemplify how it works using our example once again: suppose the Volunteer asked for the number 23.

In the first round, the Magician counts to 23, and sees in which column the 23rd card falls — it will be in the second column. This tells us that the column with the chosen card must be placed in the second position, the middle one — here the positions in the deck are as follows:

faces up	faces down
position 3	position 1
position 2	position 2
position 1	position 3

In the second cycle, the Magician counts the cards again, as he puts them on the table, and when he reaches 23, instead of saying (mentally) 23, he says 3. From then on, he only counts the cards that fall under the card 23 — that is, he will make a count every 3 cards. When a card falls in the same column as the previous one, he says 2, and then 1, continuing cyclically: 3, 2, 1, 3, 2, 1,... The last number you say is the position in the deck where you put the column with the chosen card. In our case, the 23rd card falls on the 8th row, so the Magician just says, "3, 2," knowing that he has to put, again, the column with the chosen card on the 2nd position of the deck.

In the third cycle, there is no counting, just see if the number is between 1 and 9, 10 and 18 or 19 and 27. In the first case, it will go to position 1, in the second to position 2 and in the third to position 3 — in our case it will be 3, on top of the other two columns, before turning the deck over.

The process of collecting the cards, in the first effect, can raise suspicions because it is not always done in the same way. Magician Dai Vernon has suggested a method to disguise this difference. When collecting the piles of cards, we can drag them, with the right hand, to the edge of the table next to us, letting them fall onto the left hand. We always do this from left to right. To put a column underneath, or in the middle of the others, you have to maneuver the piles already collected with the fingers of your left hand.

Mathematics

As with the previous effect, you need to see how the cards circulate each time they are collected and redistributed.

Let's separate the set of cards, organized in three columns on the table, into blocks numbered 0 to 2, with three lines each — for technical reasons, it is always convenient to start the numbering at zero when doing the mathematical analysis of this effect, which will cause a discrepancy of one unit in some counts.

We number the rows, within the blocks, also from 0 to 2 and the columns in the same way.

	col. 0	col. 1	col. 2	
	X	X	X	row 0
Block 0	X	X	X	row 1
	X	X	X	row 2
	X	X	X	row 0
Block 1	X	X	X	row 1
	X	X	X	row 2
	X	X	X	row 0
Block 2	X	X	X	row 1
	X	X	X	row 2

We will also number the position of the columns in the deck after the cards are collected:

faces up	faces down
position 2	position 0
position 1	position 1
position 0	position 2

There are three principles that describe the movement of cards from one position on the table to the next (which is obtained by collecting the piles and re-distributing the cards). We will represent by i a number that can be 0, 1, or 2. From one cycle to the next the following happens:

- The cards from the column at position i in the deck move to block i.
- The cards from block i move to rows i (regardless of how the cards are collected).
- The cards from row i move to column i (regardless of how the cards are collected).

As with the previous trick, these principles can be confirmed through experimentation. You can use, for example, red cards for the set in question and black cards for the remaining positions. We will now number the column positions from 0 to 26 using base 3, according to the order in which the cards are laid out on the table. This order is also the order of the cards in the deck (face down):

$$
\begin{array}{ccc}
000 & 001 & 002 \\
010 & 011 & 012 \\
020 & 021 & 022 \\
100 & 101 & 102 \\
110 & 111 & 112 \\
120 & 121 & 122 \\
200 & 201 & 202 \\
210 & 211 & 212 \\
220 & 221 & 222
\end{array}
$$

Comparing this picture with the previous one, we arrive at the fourth principle:

In each position, the first digit is given by the block, the second by the row, and the third by the column.

Thus, we can conclude the following result: *Let ijk be the expression of a position in base 3. A card will find itself in position ijk after three cycles if the column it is in is placed successively in positions k, j, and i of the deck as it is collected.*

This result justifies a slightly more elaborate method than the ones we have presented. For the first effect, we need to subtract one unit and write that number in base 3. In the example above, the Volunteer chose 23. Now,

$$22 = 2 \times 9 + 1 \times 3 + 1 = 2 \times 3^2 + 1 \times 3^1 + 1 \times 3^0$$

That is, in base 3, the number 22 is written 211 (to know more about base 3 see the annex about positional numbering systems). Reading these numbers from right to left, we have $k = 1$, $j = 1$, $i = 2$. They then tell how many columns you put under the one containing the chosen card each time you collect the cards, before turning the deck face down. We then have the following practical rule:

base 3 coeff.	columns below
0	none
1	one
2	two

In our example, the first two times you have to put the column containing the chosen card in the middle, the third time on top.

For the second effect, the reasoning is similar, with the Volunteer providing the digits with his choices.

The first method we presented for the two effects is an adaptation of this one, simplifying the decomposition. As for Walker's method for the first effect, we will not describe its operation in detail, we only give the general lines of explanation of its operation. Let's call the position n, after subtracting one unit, $0 \leqslant n \leqslant 26$. The first count is a way to know which column n is in, which defines the third digit. The second is a way of knowing the number of the row in which n appears, which defines its second digit. Both counts are cleverly done in order to give the desired result, which is not difficult to confirm. In the third cycle, the three intervals correspond to the three blocks (which can easily be verified). Thus, it is again a way to find the position decomposition in base 3.

Three rows — but *not that trick!*

Effect

The Magician shows the Volunteer a magic deck, in which the cards are so closely related to each other that they are capable of knowing when one is missing. He arranges this deck on the table in three columns of five cards, gives the rest of the cards to the Volunteer, and asks her to choose one and memorize it. He then asks her to replace one of the cards on the table with the chosen card, which must be of a different color. To avoid mistakes, the Magician should demonstrate before giving the rest of the deck to the Volunteer to make the switch. The Volunteer can make the switch in the Magician's sight, or without the Magician seeing — in that case, after the switch, she must close the columns of cards into three piles.

Now, the Magician collects the 15 cards and places them on the table, again arranged in three columns, turning them over one by one, just as in the "three times seven, 21" trick. At this point the Volunteer may think she already knows the trick, and that the Magician will ask her in which column the card is. However, the Magician insists, as he places the cards, that even though the column arrangement may seem familiar to the Volunteer, "this is not that trick": There is no need to point to any column; the Volunteer just has to concentrate on the chosen card.

The Magician takes the cards and puts them back on the table. He insists that this is not the usual trick, and that no information is needed from the Volunteer, just that she should concentrate on the card. He again collects the cards in the same way, and again lays them out on the table in the same way, again asking that this time the Volunteer concentrate really hard. At this point he calls a Helper, who has seen nothing of the process, and asks him which card he has chosen, which he identifies (if the Volunteer has made the initial switch without the Magician seeing anything, there is no need for a Helper, the Magician can guess).

As a variant, the Magician can split the initial magic deck with the Volunteer, asking her to choose 12 cards, to which he adds 3 cards chosen by him.

Method

The method of collecting the columns must always be the same: the Magician collects the cards, forming three piles face up, and joins them by placing the rightmost pile on the middle one, and both on the left. He then turns the pack upside down and arranges the cards from left to right, turning them over successively to form three columns.

The method of discovering the card is based on an error-detecting code: a set of four words each made up of five symbols, where each symbol can have the value "R = red" or "B = black". The words are:

$$BBBBB \quad BRRBR \quad RBRRB \quad RRBRR$$

These words are formed in the following way: If a and b are any two colors, the pattern is a, b, $a+b$, a, b — for the sum, assuming that B is worth 0, R is worth 1, and $1+1=0$. Thus, each pair (a,b) of colors generates a word.

This code has the following property: If we change the value of any of the symbols in a word, it is possible to detect which symbol was changed.

For example, if we are presented with the word $BRRRR$, we find that the pattern of the first two symbols is not the same as the last, so it will be one of these four that is wrong. Since $B+R=R$, it is the second pair, RR that is wrong, and therefore the first is right. Comparing each of the symbols, we see that it is the fourth R that should be B — that was the changed card.

In another example, if we are presented with the word $BRBBR$, we see that the pattern in the first symbols is the same as in the last symbols, so they are correct. Since $B+R=V$, the error occurred in the middle symbol.

We invite the reader to try some more examples and see that it is always possible to detect the altered symbol. This code also has a symmetry: if we read a valid word from right to left, it is still valid.

To prepare the deck we then need three sets of five cards, whose colors match one of the four code words. For the variant, given 12 cards, it is always possible to find three more to get three valid words (it is simple to figure out what to do in each case).

The magic deck is formed by arranging these three sets of five cards in columns on the table. Only one extra condition is required: The first card of the first pile (top left) must have the same color as the last card of the last pile.

The Volunteer, when inserting his card, changed one of the symbols of a word. If there had been no change, after arranging the cards on the table three more times (besides the initial one), each column should be a valid word. By detecting which column is wrong, and which card is wrong in that column, the Helper identifies the chosen card.

To see now what happens to the cards when we deal them on the table, suppose the cards initially (before the first deal) look like this — note that these columns do not represent code words, they only serve to mark the places at the beginning of the trick:

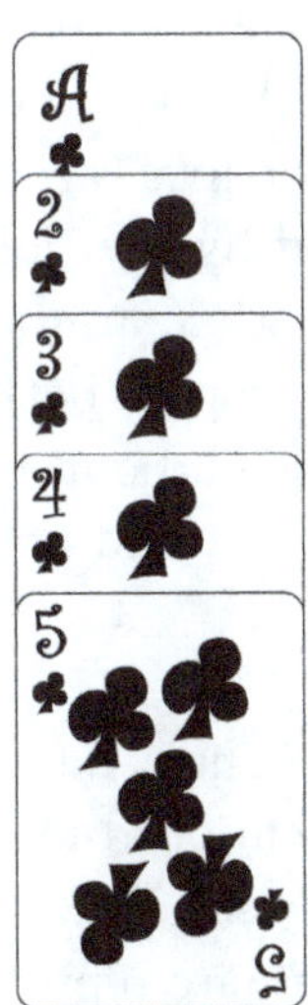

After we have collected and dealt the deck on the table three times, the positions of the cards are as follows:

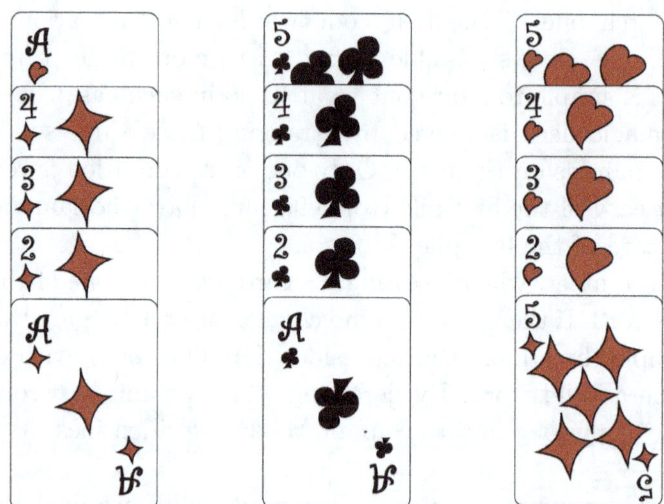

That is, the central pile of five cards only changed the direction of the cards. The other piles changed position and the cards also changed direction, except for the first and the last card, which were chosen as having the same color in our deck. So, if we start with three valid words, we end with three valid words.

Now, initially, each set of five cards was a valid word, except for the one that was changed by the Volunteer. So, only two of the vertical rows, in this final position, are valid words. It doesn't matter whether you read them from top to bottom or from bottom to top. So, all the Helper has to do is find out which is the wrong word and the wrong symbol in that word, and he will have found the chosen card.

Mathematics

We have already explained how the code works when describing the method. This effect is mentioned in Colm Mulcahy's *Card Colm blog*, which has other code-based effects — for example, a ternary code where the three numbers are represented by black, red or face-down cards.

Memory of colors

Effect

The Magician asks the Volunteer to choose five cards (or 10, or 15, or 20...). The Magician arranges these cards on the table, saying that he has achieved a perfect balance of colors, so that if any color is changed, the balance will be altered. He then asks the Volunteer to change the color of one card (replacing it with a card of another color). After the change is made, the Magician calls a Helper, who has seen nothing of the process, and who, looking at the table, says which card has been replaced.

Method

Let's describe the method for five cards. For 10, 15 or 20, we apply this method to each group of five.

The Magician must arrange the cards in a precise manner on the table. Here again, the code mentioned in the "three rows" trick is used. Remember that the valid words in this code are 00000, 01101, 10110, and 11011. The rule is: The first two symbols are repeated at the end of the word, and the middle symbol is the sum of these two numbers, with the convention $1 + 1 = 0$. With this code, you can detect any symbol change in a valid word.

With the five cards the Volunteer chooses it is always possible to make a valid word in this code, as long as you assign appropriate color values to the numbers 1 and 0. It is not difficult to decide how to make the assignment, anyway, here is a table.

Pattern	Values	Words
$BBBBB$	$B = 0, R = 1$	$BBBBB$
$BBBBR$	$B = 1, R = 0$	$BBRBB$
$BBBRR$	$B = 1, R = 0$	$BRBBR, RBBRB$
$BBRRR$	$B = 0, R = 1$	$BRRBR, RBRRB$
$BRRRR$	$B = 0, R = 1$	$RRBRR$
$RRRRR$	$B = 1, R = 0$	$RRRRR$

Given the colors he has, the Magician decides how to assign values to the colors, and codifies this choice by the order in which he puts the cards of each color: $B = 0, R = 1$ (Increasing order); $B = 1, R = 0$ (Descending order).

We then order the cards within each suit from deuce to Ace. Among the suits, we can choose the CHaSeD order, but considering the separate colors: clubs < spades in black and hearts < diamonds in red. That is: the clubs come before the spades, and within each suit, the order of values is considered. The situation is similar for hearts and diamonds.

The Magician then puts the cards on the table, so that the word is valid, and puts the black and red cards in descending or ascending order, according to the value assignment he made. One of the cards is now exchanged for a card of a different color, and the Helper has to figure out which one it is. To do this, he first detects which value assignment was made. To do this he will look at the most abundant color, that is, that color which appears on at least three cards.

- If all the cards are the same color, then the original word was surely 11011, so the selected card was the middle one.
- If four cards are of one color and one of another, the Helper looks for the predominant order in the four cards. Suppose that among these four, the second is greater than the first, the third less than the second, and the fourth less than the third. In this case the predominant order is the decreasing order and the assignment is $B = 1$, $R = 0$.
- If three cards are of one color and two of another, the Helper looks at the order of the three cards of the same color. If they are in ascending or descending order, that is the set order. If this is not the case, the order is that given by the two cards of the other color.

After figuring out which color is worth 1 and which is worth 0, it is simple to find out which card breaks the pattern. We now present two examples (see also the examples in the "Three rows" trick). Suppose the initial cards are 2♣, 4♥, 8♥, 8♠ and 3♦. We are in the case $BBRRRR$, so we make the assignment $B = 0$, $R = 1$ and the order is ascending. The Magician then chooses the word $10110 = RBRRB$. Since 2♣ < 8♠, and 4♥ < 8♥ < 3♦, the cards are in the following order on the table:

1. If the Volunteer wants, for example, to exchange the Jack of clubs for a red eight (necessarily the eight of diamonds), the four red cards are in no particular order:

However, the dominant order is increasing: the sequence grows from 4♥ to 8♦ and from 8♥ to 3♦. The value assignment is $B = 0$ and $V = 1$ and the word is 11110. Since the initial pattern is different from the final pattern, one of these must be wrong, and since the central card is $1 = 1 + 0$, the wrong pattern is the first, and the wrong card is then the second.

2. Returning to the initial set, suppose that the Volunteer wants to exchange the eight of hearts for the King of spades. We are left with the following situation:

The three black cards are in no particular order, but the two red cards are in ascending order. Again, this indicates that the value assignment is $B = 0$, $V = 1$ and the word is 10010. The pattern is the same at the beginning and the end of the word, so it must be the central symbol that is wrong, and this is confirmed because $1 + 0 = 1$ and not 0.

Mathematics

This is another trick based on the error-correcting code, coupled with the possibility of transmitting an assignment of values with the order of the cards.

This effect is mentioned in the Card Colm's column, the presentation was developed by Pedro Freitas.

One fortune serves eight

The Magician invites the Volunteer to mentally choose a card. He then gives her half of the deck and asks her if her card is there. It may or may not be, but in either case, the Magician gives the Volunteer the part of the deck containing the chosen card and keeps the other half, which he distributes in piles on the table.

The Magician then announces that, even without looking at any card, he will try to guess the card chosen by the Volunteer. He begins by lifting the various piles, one by one, with the faces of the cards facing the Volunteer, without the Volunteer seeing any cards. Each time he asks if a card of the same value as the chosen card appears in that pile, since the chosen card cannot appear because it is in the half of the deck that is with the Volunteer.

After finishing this phase, the Magician announces that he already has an idea of the card's value, but is not yet sure of its suit. He then shows the Volunteer the last pile again and asks if she sees any cards of the same suit as the chosen card.

With this, the Magician announces that he has already figured out which card was chosen because the Volunteer had to concentrate so hard on its value and suit. He then asks for the Volunteer's pack, and looking at the cards, chooses one which he places on the table, initially face down. Finally, he asks the Volunteer to say aloud what her card was, and the Magician, turning the card he selected face up, reveals that it is the one chosen by the Volunteer.

Method

The deck must be arranged as follows from the top, faces down: A♥, Q♠, 3♣, 7♥, 10♠, 8♠, 4♥, 2♣, K♣, Q♥, 5♣, 6♠, 10♥, 8♥, K♥, 3♥, 4♣, 2♥, 7♣, 9♣, A♣, J♣, 6♥, J♥, 9♥, 5♥. This is the first half. The following cards follow in the second half: A♠, Q♦, 3♠, 7♦, 10♦, 8♣, 4♦, 2♠, K♦, Q♣, 5♦, 6♦, 10♣, 8♦, K♠, 3♦, 4♠, 2♦, 7♠, 9♠, A♦, J♦, 6♣, J♠, 9♦, 5♠.

When the Magician separates the deck, it must be exactly in half. He then retains the half that doesn't have the chosen card, and deals the cards, face down, into five piles, as follows: He deals one card to each of the five piles, then a second, a third and a fourth. Now, there are five piles, each with four cards, and the Magician has six cards in his hand. He then deals one card to each of the first four piles, and the remaining two cards to the first two piles. So, these first two piles are left with six cards, the third and fourth with five, and the fifth with four.

To each of the piles corresponds a numerical value. The values are from the first to the fifth: 1, 4, 2, 7, 8. The Magician can memorize these values using the sentence *One Fortune Serves Eight*, which gives the trick its name.

When the Magician asks the Volunteer if there is a card in each pile with the same value as the chosen card, the Volunteer will answer "yes" either one or two times, regardless of the chosen card. This allows the Magician to stop showing cards after the second "yes" if he wants to.

The Magician must then add up the values corresponding to the two piles where the Volunteer said "yes," and this will be the value of the chosen card. For example, if the Volunteer says "yes" on the second and third piles, the value of the chosen card is $4 + 2 = 6$.

There is only one special case to take into account: the Jack has the value 11, the Queen 12 and the King 15, and not 13 as you might expect. Thus, if the piles chosen are the last two, the sum will be 15 and the Magician knows that the card chosen is a King.

As for the suit, when the Magician shows the last pile again, the answer can be "yes" or "no". Here, the code is as follows: "no" means that the chosen card is red, "yes" means that it is black (as a mnemonic, the Magician may think that the letter "o" of "no" has a round shape, and remember a red cherry). With this, he knows the value and color of the card, and upon examining the Volunteer's portion of the deck he will find only one card with these characteristics.

Mathematics

This trick is based on an ingenious order in the deck. By placing the cards on the table, whichever half is chosen, the piles are left with the following cards:

1st pile value = 1	2nd pile value = 2	3rd pile value = 3	4th pile value = 4	5th pile value = 5
A	Q	3	7	10
8	4	2	K	Q
5	6	10	8	K
3	4	2	7	9
A	J	6	J	
9	5			

As you can see, each card appears in only one or two columns, and adding the values of those columns gives the value of the card. For example, there are Jacks only in the second and fourth piles, and $4 + 7 = 11$.

As for the suit, let's assume that the Volunteer has the first half of the deck, which contains his card. This means that in the fifth pile are 10♦, Q♣, K♠, 9♠. If the Volunteer says "no", clearly his card is red (it is really a Heart, but that information is not necessary). If he says "yes", in principle, it could be a diamond, a spade, or a club — but it cannot be a diamond because these are all in the second half. Thus, it is a black card. In either case, an analysis of the half of the deck that is with the Volunteer settles the question. For example, if the card is a Jack, in the first half of the deck there is only one black Jack and one red Jack.

In general, this kind of tricks uses the additive decomposition of a number as a sum of powers of 2 (as in other tricks in this chapter). In this case, the additive decomposition ensures that each value is decomposed as a sum of at most two parts. This limit on the number of summands is necessary because in each half of the deck there are only two cards with a certain value.

This effect is described in the journal *Genii*.

Cheney five-card trick

Effect

A Volunteer chooses any five cards from a deck and gives them to the Magician's Helper. The Magician returns one of the cards, which the Volunteer puts in his pocket. The remaining four cards are placed side by side on the table, possibly with some face up and some face down, or handed to the Magician, who has entered the room in the meantime, but has seen nothing of the choosing process. The Magician looks at the four cards and guesses which card is in the Volunteer's pocket.

Method

In five cards there is at least one repeated suit, and the Helper gives one of the cards of that suit back to the Volunteer. He then uses the cards he has to encode the value and suit of the hidden card. There are several ways to do this, we will present nine variants. The value of the numerical cards is clear; for the face cards we'll take Jack $= 11$, Queen $= 12$, and King $= 13$.

We illustrate the explanation with an example, suppose the Volunteer draws the following five cards:

    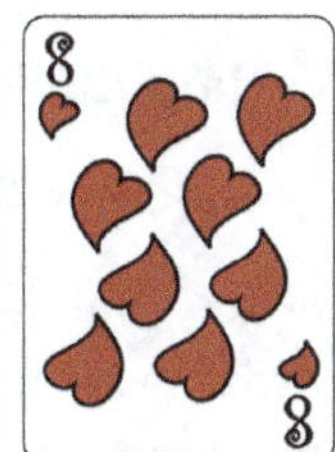

Variant 1 (Luís Sequeira): The Helper encodes the card's value in base 2 (see table in the appendix) and uses the four cards to write that value, understanding that a face-up card is worth 1 and a face-down card is worth 0. The card of the same suit as the chosen card appears as far to the left as possible. In the example, he might return the Queen of hearts or the eight of hearts — let's assume he returns the eight of hearts. Having then to encode the number 8, the arrangement for the remaining four cards would be:

If you had returned the Queen, a possible coding could be

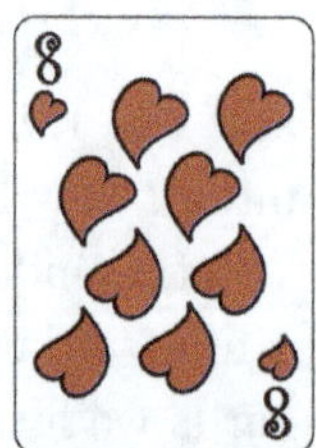

This is perhaps the simplest variant, but not the most economical, if we admit the possibility of having cards facing up and down (see variant 5).

Variant 2: Sorting the cards cyclically,

$$A < 2 < 3 < 4 < 5 < 6 < 7 < 8 < 9 < 10 < J < Q < R < A$$

we see that if we have two different cards, they will be at most six cards away from each other. Again, there have to be two cards of the same suit, let's call them X and Y. Let's consider that when going from X to Y, in ascending order, the path is shorter than when going from Y to X — in this case we say that Y is the highest card. This means that the value to add to X to get Y is less than or equal to 6. The Helper then returns this highest card Y and places the X card on the leftmost position on the table. With the remaining three he encodes, in binary, the difference between the card X (on the table) and the card Y (returned) — again, an up card is worth 1 and a down card is worth 0. In other words: the number encoded is the one that must be added to the leftmost card to get the chosen card.

Going back to the example, the Helper knows that he must choose one of the cards of repeated suit. It has to be a heart, and there is an 8 and a Queen. In the order described, the highest card is the Queen, that is, the difference between Queen and 8 is 4 (8 < 9 < 10 < J < Q), while the difference between 8 and Queen is 9 (Q < K < A < 2 < 3 < 4 < 5 < 6 < 7 < 8). So, the Helper gives the Queen of hearts to the Volunteer, who puts it in his pocket.

Of the four cards left, the Helper puts the eight of hearts on the far left, so that the Magician knows the suit of the hidden card. Three cards are left to tell which now in the cyclic ordering described above, how many units must we go up from 8 to reach the Queen? Four units, which in base 2 is written 100. So, the Helper leaves the Magician with the following configuration:

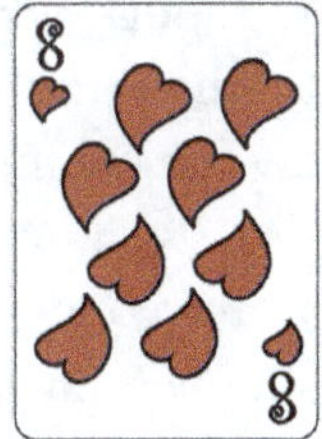   

The Magician has no difficulty in recognizing that the hidden card, being a heart, is obtained by adding 4 to 8, which gives 12, which is the order number of the Queen. If the hidden card were a 10 (instead of the Queen), the encoded number would have to be a two, which is the difference $10 - 8$. Since two is written 10 in base 2, a possible configuration would be

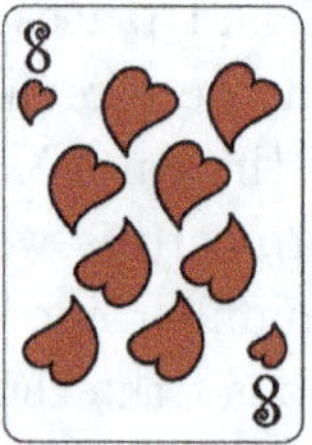

Variant 3: The main difference between this variant and the previous ones is that the four cards are placed on the table face up. As in variant 2, the Helper returns the highest card of the repeated suit (considering the cyclic order) and it is the other card of the same suit that is left on the leftmost position, face up. Now it is necessary, with the remaining three cards, to encode a number from 1 to 6, which will tell you how much you have to add to the left-hand card to get the chosen card. This is achieved, in this case, by thinking that there are exactly six different ways to line up three cards. To do this we need a complete order in the deck (we'll use a different order than the one used in the trick "Memory of colors"). We consider the cards sorted by value (a deuce will always be lower than a face card, for example, with the Ace being the lowest value card). Within cards of the same value, we use the order given by the word CHaSeD, which sorts the suits as clubs < hearts < spades < diamonds (the initials of the word refer to clubs, hearts, spades, and diamonds). For example, in our five-card set, we have: $A\clubsuit < 8\heartsuit < 8\diamondsuit < 9\spadesuit < Q\heartsuit$.

Thus, in a set of three cards, there is a first (F), a second (S), and a third (T). We encode the various orderings using the natural order of three-digit numbers: $123 < 132 < 213 < 231 < 312 < 321$.

$$FST = 1, \ FTS = 2, \ SFT = 3, \ STF = 4, \ TFS = 5, \ TSF = 6$$

In our example, after returning the Queen of hearts, the Helper must first lay out the 8 of hearts. For the remaining 3 cards, the order is $P = A\clubsuit < S = 8\diamondsuit < T = 9\spadesuit$.

Since we have to report the number 4, the order has to be STF, the cards are arranged as follows:

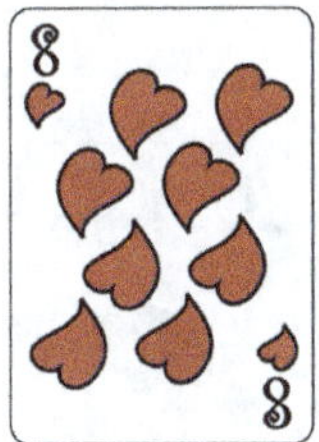 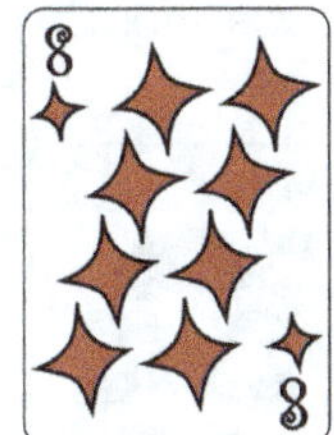  

Once again, the Magician learns that the card chosen is a heart, and that to find out its value he must add 4 to 8, yielding 12.

Variant 4 (Colin Wright): As in the previous variant, here all cards are turned face up. The Helper returns the highest card of the repeated suit, retaining the lowest card. This is placed in the second position from the left. The remaining three cards are considered in order, there being a first, a second and a third, considering the complete order of the deck described in the previous version. The convention we will adopt is as follows: The first card is worth 1 or 2, the second is worth 3 or 4, and the third is worth 5 or 6. One of these numbers, added to the value of the lowest card, will give the value of the hidden card. The card corresponding to the correct value is placed to the left of the lowest card, of the same suit as the hidden card. The remaining two cards are placed to its right. If you want to encode the lower of the two possible values, the lower card is placed next to the second card, immediately to its right. To encode the higher value, the higher of the two cards is placed in this position.

We resume the previous example, with the respective total order for the deck. We have the following cards: A♣, 8♥, 8♦, 9♠ and Q♥. The highest card of hearts is the Queen of hearts, which is returned, and you will need to add 4 to the other card of the same suit, the 8♥, to get the value of the Queen. The remaining three cards are sorted as follows, as in the previous variant: A♣ < 8♦ < 9♠. The value 4 is the highest possible value of the second card, which is 8♦.

This card is then placed to the left of the 8♥. Of the remaining two, it is the largest, 9♠, that is placed immediately to the right of the 8♥, to indicate that the largest possible value for the 8♦ should be considered. The arrangement then looks like this:

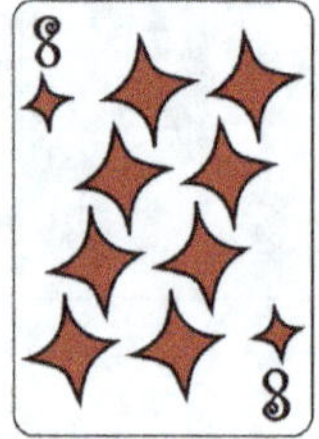   

The attentive reader will see that this encoding is similar to the previous variant, but the different interpretation given here may make it easier to read the value to be added.

Variant 5: In this variant, the Helper only has to show three cards to the Magician for him to guess the chosen card. In a combination of Variants 1 and 2, the Helper returns the highest card of the repeated suit (considering the cyclic order) and encodes the number that must be added to the lowest, writing this number in binary. Since there must be at least one card face up, the Helper places, in the leftmost position, the lowest card face up. Again, in our example, the arrangement would be

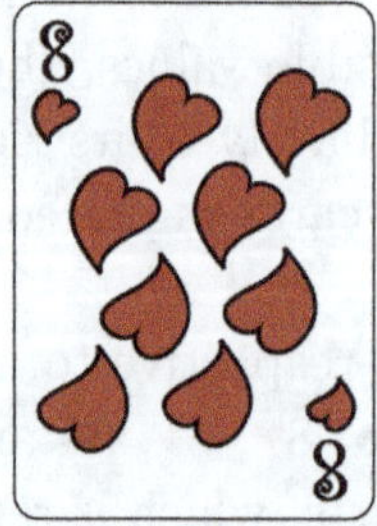  

The next two variants have a more laborious method. Their interest is largely theoretical: In Variant 6, a way of doing the trick with four cards instead of five is presented, and in Variant 7, the

Volunteer is given the possibility of choosing the card from the five chosen initially. Before reading the descriptions, you might want to think about these problems, to see if you can find a method for each of these variants from the techniques already presented.

Variant 6 (Colm Mulcahy): The Volunteer selects four rather than five cards, of which the Helper returns one to him. The way to encode the chosen card is similar to Variant 4. The problem is that in this case we are not sure that there is a repeated suit. To circumvent this, we use the concept of augmented suits, which requires some training. We will then have three augmented suits, A, B and C, instead of the usual four, with the cards ordered as follows:

A: all clubs, in order, followed by 2, 3, 4, and 5 of diamonds; B: all hearts, in order, followed by 6, 7, 8, and 9 of diamonds; C: all spades, in order, followed by 10, J, Q, and K of diamonds.

The only card that was not included in any augmented suit was the Ace of diamonds, which will be a special card. Each augmented suit then has 17 cards, which, as in Variant 2, we will consider cyclically ordered. We now follow the steps of Variant 2: Of the three cards chosen we have to have two cards in the same augmented suit, and furthermore, in cyclic order, there has to be a difference between these two cards that is less than or equal to 6. Thus, there is a highest card and a lowest card, and the Helper returns the highest card to the Volunteer.

If the chosen card is the Ace of diamonds, the Volunteer sets all cards face down. Otherwise, the Helper places, as in Variant 5, the lowest card in the left position, face up, and encodes in binary the difference between this and the chosen card, if this difference is a number between 1 and 6. For the numbers 7 and 8, all cards are face up, and more information needs to be conveyed: Considering the complete order in the deck, the two cards that appear to the right of the lowest card may be in ascending order, for the 7, or descending order, for the 8.

Variant 7 (Victor Eigen and Martin Gardner): Suppose we want to let the Volunteer choose the card, out of the initial five. How can the Helper communicate the chosen card to the Magician?

For this purpose, you must use the technique of Variant 3, ordering all the cards in the deck and considering that there are 24 permutations of a set with 4 elements. Now, in the deck, after removing four cards, there are 48 cards. The Helper then encodes a number from 1 to 24 using the permutations, and gives the cards to the Magician either face up or face down, thus indicating whether or not he should add 24 to the encoded number. Alternatively, if instead of giving him the cards in hand, he puts them on the table, he can start from left to right or right to left, with the Magician already present in the room, of course.

All the numbers from 1 to 48 can thus be encoded. Thus, the main difficulty in the method is that the numbering of the cards is not completely fixed because when the Magician receives the four cards, he has to remove them from the order of the deck, and renumber all those that come after them.

The next variants present another possible answer to both previous restrictions: they can take place with four or five cards, and the Volunteer has the right to choose which card he gets.

They are based on ideas of Tiago Hirth, and their implementation only requires that cards may be placed on the table in a horizontal position in addition to the vertical one.

Variant 8: The Helper asks the Volunteer to choose five cards, and from these choose one and put it away. The Helper now has four cards and will place them all face up on the table, codifying six numbers, all equal to 0 or 1. Having established a total order in the deck, the Helper then begins by separating the four cards into two groups, and orders the cards in each of the two groups, agreeing that the increasing order is worth 0 and the decreasing order is worth 1. These two numbers will codify the suit, agreeing for example clubs $= 00$, hearts $= 01$, spades $= 10$, diamonds $= 11$.

This then defines the order of the four cards on the table. To give the value of a card, the Helper writes it in binary (as in Variant 1), agreeing that Vertical $= 1$ and Horizontal $= 0$.

Let's go back to our example, and suppose we set the order of the cards in the deck as in Variant 3. Suppose that the Volunteer chooses the Queen of hearts again. We then have to put on the table an increasing sequence of two cards followed by a decreasing sequence, using the cards we have, thus giving the suit information, 01 = hearts. Since 12, which is the value of the Queen, is written in base 2 as 1100, the final arrangement looks like this.

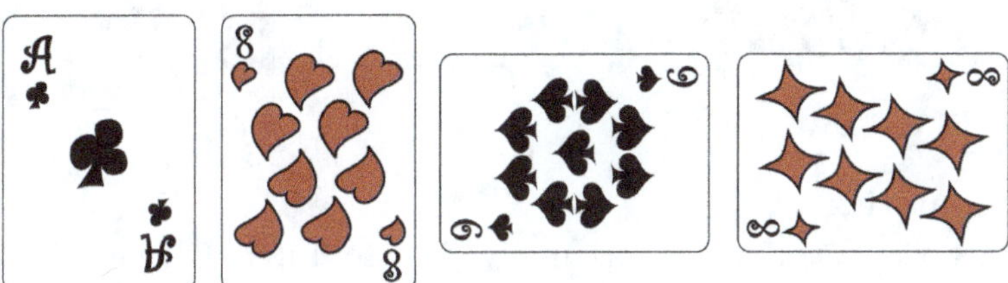

Variant 9: In this variant, the Volunteer chooses four cards and is entitled to hold one, known to the Helper. The Helper then arranges the remaining three cards on the table, and is free to place them either up (U) or down (D), and horizontally (H) or vertically (V). This enables four combinations, in which a number from 0 to 3 can be encoded, for example

$$0 = D + V, \quad 1 = D + H, \quad 2 = U + V, \quad 3 = U + H$$

The Helper uses the leftmost position to convey the suit of the chosen card, agreeing upon an order for the suits is established that naturally associates them with a number from 0 to 3. The CHaSeD order suggests clubs $= 0$, hearts $= 1$, spades $= 2$, and diamonds $= 3$.

The two remaining cards give the value of the chosen card, in base 4. In this base, the digit in the leftmost position is multiplied by 4. Thus, if cards are laid on the table with the values X Y, where X, Y $= 0$, 1, 2 or 3, the number transmitted is $4X+Y$. With this choice, all numbers up to 15 can be encoded, which is sufficient for the values of the cards.

For example, excluding the eight of hearts from the initial five cards, if the Volunteer chooses to keep the Queen of hearts, we will have to arrange the remaining cards on the table as follows:

Mathematics

This wonderful trick, originally due to Fitch Cheney Jr. and published by W. Wallace Lee in 1950, uses several mathematical concepts in a very ingenious way.

An argument presented in the first five variants guarantees that in five cards, there are at least two of the same suit. This is the Pigeonhole Principle, which, besides being used in sophisticated demonstrations in mathematics, also has funny applications. One of its possible statements is: *If we distribute $n + 1$ letters among n mailboxes, then at least one mailbox will receive more than one letter.*

If we imagine four boxes, each representing a suit, we see that five cards must necessarily show at least one repetition.

A related trick question: Are there two people in Lisbon with exactly the same number of hairs? The Pigeonhole Principle allows us to answer in the affirmative, since nobody has more than 400,000 hairs and Lisbon has one million inhabitants.

In the variants that use base 2 numbering, it is simple to see that with 4 binary digits you can write any number from 0 to 15, and with 3, any number from 0 to 7. Also, in base 4, with 2 digits you can write any number from 0 to 15.

In the other way of encoding, we use the number of possible permutations of three objects, which are six, or of four objects, which are 24, allowing us to write any number from 1 to 6 or from 1 to 24, respectively, in general, the number of permutations of n different objects is

$$n! = n \times (n - 1) \times \cdots \times 3 \times 2 \times 1$$

In either case, it is possible to transmit the desired information.

Gilbreath Galore

This chapter has several effects based on Gilbreath's two beautiful principles, which we will explain as they are needed.

One black, one red

Effect

The Magician shuffles the cards once (or asks a Volunteer to do so) and places them behind his back. He successively removes a pair of cards of different colors.

Method

You prepare the deck, so that the colors of the cards alternate.

The deck must be broken into two piles, so that the colors of the bottom cards in each are different. Shuffle these two piles once with a *riffle shuffle*.

If the Volunteer is the one doing the shuffling, the Magician should look at the colors of the top and bottom cards when he receives the deck. If he notes that these cards have different colors, he should cut the deck between two cards of the same color.

If we successively draw pairs of cards from the top (or bottom) of the deck, every pair will have a card of each color.

Variant: Dai Vernon has created a slight variation on the method of presenting the cards that allows the black cards to be placed in one pile and the red cards in another, as they come out of the deck. To do this, the Magician holds the deck in his left hand, with the cards face up. He then places his hands behind his back, or under the table, and removes the two cards from the top, holding one in his left hand and another in his right hand. He must place the one in your left hand behind the deck, turning it around, facing away from the rest of the cards.

He sets these two cards on the table, starting with a red and a black pile — but also take a quick look at the color of the card that is now in view, so he knows whether to put it in his left hand or his right hand to match the colors of the piles. The next card goes to the other hand.

Mathematics

The justification can be formalized in a demonstration by induction and involves noting that the two cards at the bottom of the deck came from the same pile (and are of different colors because that's how we got them) or come one from each pile (these colors are also different). We conclude that, in any case, the two bottom cards have different colors. We can then move on to analyze the next pair, and analogous reasoning allows us to conclude that these too have different colors, and so on.

Now, let's see what happens if the shuffle is done wrong. If the bottom cards of the two piles that are shuffled have the same color, then the top cards also have the same color, and that color is distinct from the color of the bottom cards. Therefore, in the shuffled pile, the bottom card has a different color than the top card (note that if you shuffle properly, you can have both coinciding and opposing colors of the bottom and top cards, but if the shuffle is done badly, the cards will necessarily have different colors).

We then propose that the deck be cut between two cards of the same color. If the shuffle has been done well, two cards of the same color are necessarily in different pairs, and the cut maintains the desired structure.

If the shuffle was done incorrectly, the order is correct if you pass one card from bottom to top (this move, made before the shuffle, would have resulted in two cards of different colors underneath). Now, a cut between two cards of the same color can be thought of as the passing of a card, which corrects the situation, followed by a cut, which does not change the structure. Of course, it would be enough to pass the card from the bottom to the top for everything to be fine, but the cut produces a better effect.

This famous principle is due to Norman Gilbreath, who was the first person to use it to produce magical effects. The description appears in the article *Magnetic Colors* from 1958. It is known as Gilbreath's first principle (we will talk about the second one soon).

The swapped pair

Effect

The Magician presents a small deck of cards which he shuffles and places on the table in piles of two cards, placing in each pile one card down and one card up. Turning his back, he asks the Volunteer to choose two cards of different colors, swap them, and turn their piles upside down to hide them. He can also flip over other piles, as many as you like.

When finished, the Magician gathers all the piles into one, with some cards down and some up. When he opens the cards in a fan, however, he notes that on one side all the visible cards are black, except one, which is one of the chosen cards. When he turns the deck over to the other side, something similar happens: All the cards are red, except one, which is the other chosen card.

Method

The starting pack must have an even number of cards, between 20 and 30, and must be arranged with alternating colors. The shuffle must be a riffle shuffle, and the bottom cards of the two shuffling parts must have different colors, just like in the previous trick.

The piles are made sequentially. By Gilbreath's first principle, each pile will have one red and one black card (one facing down and one facing up).

When the Magician collects the cards he must make a pile with the pairs that have red cards on top and another with the rest. When you put these two piles together in a pack, you only have to turn one of the piles upside down.

Mathematics

This is an effect achieved from Gilbreath's first principle. The only pairs that do not have two different colors are the ones containing the chosen cards, hence they stand out at the end.

This effect is mentioned in MacTier (2000).

Broken Gilbreath

Effect

The Magician presents a deck, which he invites the Volunteer to cut several times. He then separates it into two parts, one for himself and one for the Volunteer, both facing up. These two parts are then cut again, but each person can only make cuts on the part he or she has chosen. After the cuts, one of the piles is turned over and the cards are shuffled with a riffle shuffle into a single pile, from which the Magician draws cards in pairs. If both cards in the pair are face up, or both are face down, that pair is discarded (with the cards facing down). If there is one card up and one card down, the Magician considers the color of the face-up card. If it is red, he places the pair in one pile, if it is black, he places it in another.

At the end, there is a pile of discarded cards, a pile in which the visible cards are black, and another pile in which the visible cards are red. The Magician then reveals that in the pile where the visible cards are red, the remaining cards are also red, and so is the pile of black cards.

Method

The deck needs to be prepared with alternating colors. When breaking into two piles, they should be put on the table with the cards face up and the top cards should be of the same color. The Magician must then break his pile, so that his visible card is a different color than the Volunteer's pile. The rest of the trick is automatic.

Mathematics

It is once again the Gilbreath principle in action, in this case producing the opposite effect than usual. The initial cut ensures that there are an even number of cards in each pile, and that even if you cut, the black/red structure remains the same. With the cuts, the top cards become different colors, which forces the bottom cards to be different colors as well. When you flip over one of the piles, the bottom cards become the same color. So, when a pair of cards comes from one of the piles, the pair has different colors and the cards face the same way. However, when they come from different piles, they come one up and one down, and they are the same color! This is explained by the reasoning given in the "one black, one red" trick.

This effect is described on the YouTube channel Mismag822 (n.d.).

A pile, a number, and a color

Effect

The Magician shows a deck, which he can cut, makes a riffle shuffle in it, and separates it into two piles. He then asks the Volunteer to choose a pile, a color (red or black) and a number up to 10. Suppose she chooses the color black and the number 7. The Magician asks her to hold the pile she chose, facing her, and memorize the seventh black card, while the Magician takes the other pile.

The Magician then puts the two piles together and reshuffles them. He then takes the cards and looks for the chosen card, which he presents.

Method

There are two ways to perform this trick, depending on how you separate the deck into two piles.

The deck must be prepared by alternating red and black cards. When doing the riffle shuffle, one has to assure that the bottom cards of the two piles are of different colors. By Gilbreath's first principle, each pair of cards (from the top) has one red card and one black card.

Variant 1: The Magician separates the deck into two by dealing one card, alternately, to each pile. While the Volunteer sees which is the seventh black card in his pile, the Magician sees which is the seventh red card in his. He then puts the two piles together and makes a few cuts. He then looks for his card, holding the deck facing him. There are 25 cards between his card and the chosen card, so finding it is simple.

Variant 2: The disadvantage of the previous variant is that it takes a long time because of the separation into two piles. In this variant, the Magician cuts the deck into two piles of 26 cards each, compares them on the table, presses them together, to make sure they are equal. If any cards need to be moved from one pile to the other, they must be taken from the top of one pile and placed underneath the other.

While the Volunteer searches for the seventh black card, the Magician takes his pile and searches for the 14th card ($14 = 2 \times 7$). This is the second card of the seventh pair. After gathering the two piles and making the cut shuffle in his hand, he turns the deck toward himself and looks for the memorized card. If it is in the second part of the deck, he make another cut, so that it is in the first part (closest to the Magician). Next, he counts 24 cards and looks at the next two cards. Since there must be 12 pairs between the pair of the chosen card and the pair of the memorized card, the chosen card is the red card from that pair (which must have a black card and a red card).

Mathematics

It is another application of Gilbreath's first principle, which we have already described. The effect is mentioned in MacTier (2000).

Fourteen cards

The Magician begins by shuffling the cards in front of the Volunteer. He then removes cards from the top of the deck, one by one, into a pile on the table, asking the Volunteer to tell him to stop when she thinks this pile has about half the cards.

The Magician then riffle shuffles the two packs and states that he has never had much luck with the number 13. He then removes, one by one, 14 cards to the table (face down), showing the 14th to the Volunteer and asking him to memorize it. Next, he asks the Volunteer to shuffle the small pile of 14 cards. Once shuffled, he takes it, looks at the cards, asks the Volunteer a question, and guess which card came out!

Method

The deck must first be prepared. The cards are separated into four groups of 13, in which each group has an Ace, a King, and so on (suits can and should be mixed). Now, put the 13 cards in each group in a certain order, the same for everyone — it can be a progression order, or Eight Kings, or any other order. At the end, the piles are put together, one on top of the other. Thus, each of the four blocks of 13 cards, starting at the top, is ordered in the same way.

The first shuffle is a cut shuffle, which does not change this situation: Each of the four blocks of 13 cards, starting from the top, is ordered in the same way.

After the riffle shuffle, what happens is that each of the four blocks of 13 cards continues to contain one card of each value. When you draw 14 cards, the 14th is the only repeated card. The Magician then looks for this repeated card and asks a question that allows him to distinguish between the two: For example, if they are the nine of hearts and the nine of clubs, he might ask, "Was your card red?" If they are, for example, the two of clubs and the two of spades, the question has to be about the suit.

Mathematics

This trick is based on Gilbreath's second principle, which we now describe. Suppose we have a pattern that repeats itself in the deck, for example, A, 2, 3, ..., J, Q, K, A 2, 3, ...

The principle states that if we remove a pack of cards from the top of the deck, reverse their order, and riffle shuffle the two packs, then the first 13 cards contain exactly one Ace, one deuce, one three, and so on. The same is true for the second group of 13 cards, the third and the fourth. The initial order of the deck will disappear in this process.

The demonstration follows the same lines as Gilbreath's first principle: If you look at the bottom 13 cards in each of the initial packs, you see that the card values are the same but in reverse order. So, if you study the situation a bit, you see that the bottom 13 cards from the pack that you get after shuffling are all different. If you now remove those 13 cards, the two packs that are left have the same initial property, and you can repeat the reasoning. In Diaconis and Graham (2015), there is much more information about the mathematics of this principle. This effect is described in MacTier (2000).

Seven days a week

Effect

The Magician presents four Volunteers with a deck of cards, which he announces contains all the cards from Ace to seven of the four suits. He cuts it in half and shuffles it, then deals seven cards each. Next, the Volunteers are asked to look at the cards they've been dealt. He then states that each card represents a day of the week, with the Ace representing Sunday and the seven representing Saturday. He announces that since this deck is magical, each person has one and only one repeated card, which represents the most auspicious day for said person. The Magician then successively guesses what these days are, asking the Volunteers to confirm his statements.

Method

The deck must be prepared as follows, top to bottom, with the cards facing down: A, 2, 3, 4, 5, 6, 7, A, 2, 3, 4, 5, 6, 7, A, 7, 6, 5, 4, 3, 2, A, 7, 6, 5, 4, 3, 2.

With the deck face down, the top card should be an Ace and the bottom card a 2. The suits can be mixed, which is convenient.

The cut must be made exactly in the middle (to ensure this, the Magician can look for the Ace that is followed by a 7 in the center of the deck). When the Magician shuffles, he must ensure that no seven consecutive cards from a pack are left together (in which case the trick may fail).

The first Volunteer has a repeated Ace, which the Magician guesses (saying that Sunday is his most favourable day). He then asks this Volunteer for her cards and she shows them to him, confirming that there are two Aces and no other repeated cards. Also, he notes which (only) card is missing between Ace and 7: That is the repeat card of the second Volunteer. He places the first Volunteer's cards on the table, face down, and asks the second Volunteer's cards to confirm the prediction, shows them, notes the missing card, and places them on the table, face down over the first Volunteer's cards. The remaining repeated cards are guessed in the same way.

Mathematics

This is another application of Gilbreath's second principle, described in the "14-card" trick. In this case, in the top seven cards of the packs, there is a repeating Ace, and the remaining cards are in reverse order, ensuring that there are no repeats. After drawing from the two packs the seven cards that were dealt to the first Volunteer, the situation is repeated: The top card is the same, and the following cards are in reverse order. Furthermore, the top card is the one missing from the first seven. For example, if the cards dealt to the first Volunteer came two from the first pile and five from the second, the situation looks like in the following diagram:

$$A, 2|\ 3, 4, 5, 6, 7, A, 2, 3, 4, 5, 6, 7$$
$$A, 7, 6, 5, 4|\ 3, 2, A, 7, 6, 5, 4, 3, 2$$

Thus, the three is the top card in the remaining packs, is also the one missing from the first seven, and will be the one repeated in the second Volunteer's cards.

This effect is based on a version that uses five-card sequences, described in MacTier (2000).

The phone number

Effect

The Magician asks the Volunteer to split a deck, without completing the cut, turn one of the piles upside down, and shuffle it once (riffle shuffle). Now, draw nine cards in a row and turn some over, so that they all face the same way. Placing the nine cards on the table, the Magician reads them as a telephone number and asks someone to call that number — and it's the Magician's phone that rings!

Method

This is another application of the Gilbreath principle. The preparation is as follows: Two piles of cards are made, ordered according to the telephone number, and one of the piles is placed on top and the other on the bottom of the deck. Both piles are arranged in the same order: If the number in question is 123456789, for example, the Ace is on the bottom and the nine on top, with the piles facing upwards.

After the cutting and shuffling, the Magician removes the nine cards from the top containing the prepared cards, one by one (reversing the order). Gilbreath's second principle ensures that the nine cards drawn have exactly the intended numbers but with the order changed.

The most delicate part of this effect is the action of turning the cards over. Suppose that, having all nine cards in hand, the Magician notes that there are two face up, then three face down, then one face up, and so on. Holding the cards in his left hand, he slides the two face-up cards into his right hand, turns them together, and joins them with the three face down cards to form a group of five. This group is now turned en bloc, once again face up, and grouped with the next card. Continuing in this fashion until all the cards are facing the same way, they are put in order.

If the Magician's phone number has three equal digits, you can't do the effect this way (because a deck only has four cards of each value), you have to make an adaptation. The number that appears could be for example 31415927, the first digits of pi.

Mathematics

When you apply Gilbreath's principle in this way, within the nine cards that appear, those facing up are in order and so are those facing down. What the flipping method does is group each of these groups together, making the order total.

The four magic cards

Effect

The Magician presents a deck, briefly showing the cards to confirm that they have no special order. Outside this deck are four separate cards — the eight of hearts, the Queen of clubs, the four of clubs, and the seven of spades — each of which has a small piece of paper with a prediction on it (it could be a post-it, for example). The predictions should not be in plain sight.

The Magician notes that there are 48 cards in the deck and will ask the Volunteer to help him first select 24 and then 12 cards from that deck. To do this he asks her to remove a small pile from the top of the deck (less than half, about a third), taking the remaining 24 cards from the top of the deck (she can remove them one by one or count 24 and remove them en bloc). The Magician now says that he will shuffle those 24 cards, asking the Volunteer to break the small deck. Once broken, he turns one of the halves upside down and shuffles the two piles with a riffle shuffle, or even a rosetta shuffle (which is better, since the number of cards is small). Then, he removes the top 12 cards, showing that there are cards facing up and down, so that it can be seen that the shuffle has actually mixed the cards. He places the cards on the table, all face up, and notes that these cards appeared completely randomly because of the cuts and the shuffle.

At this point, he shows the predictions one by one: The eight of hearts says, "There are eight red cards," the Queen of clubs says, "There are three face cards," and the four of clubs says, "There are four even cards." The Magician is curious about the prediction of the seven of spades, since we already know how many black cards, number cards and odd cards there are. Looking at the prediciton, it says, "There is a red seven." All the predictions turn out to be correct.

Method

The Magician begins by drawing from the deck the eight of hearts, the four of clubs, the Queen of clubs, and the seven of spades, which will be the cards that make the predictions.

With the remaining cards, the Magician arranges three piles of 12 cards in the following pattern, from bottom to top, with the cards facing you. The odd red card under two of these piles (face up) must be a seven:

Red/Black	R	R	B	R	R	B	R	R	B	R	R	P
even/odd/court	o	e	e	c	o	o	e	c	o	o	e	c

Keeping the cards face up, the Magician then puts the two piles that have the red sevens underneath, one on top of the other. He puts those on top of the third pile, and finally, this whole set on top of the remaining cards in any order. From here on, the trick is automatic, provided the Volunteer draws at least 13 cards in the initial cut he makes (13 cards is a quarter of the deck plus one card, if the Volunteer draws about a third, this is guaranteed).

Mathematics

This is an ingenious application of Gilbreath's second principle: In the final part of the deck, each group of 12 cards has eight red cards, three picture cards, five odd cards, and a red seven, always in the same positions — the top pack doesn't need a seven because if the Volunteer draws 13 cards, that first card is removed.

This effect is inspired by another described in MacTier (2000), where this cyclical two red/one black order is presented in conjunction with three number cards/one figure, to which we add a parity pattern and the sevens. The original pattern appears in a 1987 book by Max Maven, who calls it a "schizoid rosary", perhaps because of the similarity to the arrangement of beads in a rosary, in this case with two imbedded patterns. The familiar name we give to this trick is *Gilbreath on steroids*, given the amount of predictions that can be made.

The Diaconis poker

Effect

The Magician proposes to a Volunteer to teach him how to deal the cards in the poker game in order to cheat. He begins by informing her of the value of the hands: pair, two pair, three of a kind, straight, flush, full house, poker, straight flush, royal flush.

The Magician has to clearly mention that the flush (all cards of the same suit) is stronger than the straight. He can eventually exemplify with some cards, dealing some hands. He must also explain how the dealing is done — one card for each player at a time.

When the Volunteer is ready to start, the Magician gives her a deck of 25 cards, which is simpler to start with. He starts by asking the Volunteer to deal about half the deck (it has to be at least five and at most 20 cards) and shuffle (riffle shuffle).

Next, he asks her to deal the cards to five players. After the hands are on the table, the Magician opens one or two hands to see what they are, then puts them back down. He then tells the Volunteer to collect the hands, in any order, but without mixing up the cards. He then proposes the following: The Volunteer now has a partner at the table, who is in first position (the one who receives the cards first). In the fifth position (the last one to receive the cards), there is a "victim": A high roller who must be dealt a good hand, provided that our partner is dealt a better hand.

The Volunteer deals the cards again, and the Magician turns over the fifth player's hand, showing a straight (10-Ace). After commenting that this is a very good hand, he then turns to his partner's hand (the first to receive the cards), revealing a flush! He comments that the Volunteer has a natural knack for this.

Method

We start with a deck of 25 cards arranged as follows: S S S S S S 10 V D R A 10 V D R A 10 V D R A S S S S S, where S represents a spade of some kind. In the middle appear three sequences 10-Ace, where the suits must be mixed. These cards must be on top of the full deck (we consider the deck with the cards face down). To illustrate, the Magician turns the deck face up, using the first cards so as not to disturb the order of those that are prepared. When the trick begins, the Magician turns the deck face down and selects the top 25 cards — he can take them out en bloc or deal them one by one to the table (the order is reversed, but that makes no difference). From then on the trick is automatic.

Mathematics

This effect is described in Diaconis and Graham (2015) as an application of Gilbreath's second principle.

We see that after the first shuffle, the bottom five cards are spades and the top five cards are a straight (unless the shuffle is done very clumsily and puts a spade on the first five cards, but this generally doesn't happen). Thus, the last cards dealt (which are on top of the others) are spades and the first cards (which are below) are a sequence. After collecting and dealing again, the last player has a straight and the first a flush of spades.

Order in the Ranks!

In this chapter, we present tricks that depend on having the deck prepared, with the cards arranged in a special way.

You choose

Effect

The Magician has eight cards in his hand, face down, which he can shuffle in front of the Volunteer, successively passing some from top to bottom. The Magician separates the top two cards from the deck and asks the Volunteer to choose one. The Volunteer chooses and the Magician places that card in front of her, face down, passing the other card down from the deck. The Magician does the same with the next two cards, putting the second chosen card on top of the one already there. Now, complaining that he has no card, he holds up the two top cards and asks the Volunteer to choose one to give to him (the Magician). He keeps the chosen card, places the other one underneath the deck, and repeats this once more, so that he has two cards as well (which remain face down).

The Magician repeats this process two more times: The Volunteer selects one more card for her, and then one for the Magician. Since there are now only two cards left in the deck, the Magician deals the top card to the Volunteer and takes the bottom one.

Before turning the piles over, the Magician reminds the Volunteer that he has made six choices. Commenting, "You like the red cards, don't you?", he turns over both piles and shows that the Volunteer only has red cards, while the Magician only has black.

Method

All that is required is that in the starting deck (face down) the top four cards are red and the bottom four are black. When shuffling, you must always pass at least four cards from top to bottom, taking them one by one from the top (reversing the order), which ensures that the colors remain separate (although you must shuffle an even number of times to ensure that the red ones are on top).

Mathematics

The trick is based on a principle proposed by Bill Simon. Since the first four cards are red, the first two times cards are dealt, the Volunteer is dealt two red cards and there are two red cards that go underneath the deck. Similarly, the next two times, the top four cards are black, so the Magician gets two black cards and there are two black cards that go underneath the deck. The deck now has two red cards on top and two black cards at the bottom, so the next card to be dealt is red and the next is black. The deck is now left with one red card on top and one black card on the bottom, which are dealt to the corresponding piles.

Curious variants of this trick can be done, using more cards, although the trick can become a bit time-consuming. The number of cards in the deck must always be a power of 2 — 8, 16, 32, ... We invite the reader to try to figure out what would be the right way of dealing.

Another variation is to ask the Volunteer to select eight cards of her choice from the deck. The Magician, looking at the choice, chooses four cards to put on top and makes a prediction about which cards the Volunteer will get, writing it on a piece of paper, for example. In any case, in eight cards magic, it is generally simple to choose four cards that stand out from the others: They can be court cards, or all of the same suit, or all of two particular suits, which can be announced orally.

Still another variant, this one more spectacular, is to simulate a poker game. You announce that you are going to play a game where each player has four cards and there is one card on the table that will be shared by both players. To prepare the deck, you choose five cards that make a good hand (four Aces and one more, for example), put one of them on the table, and choose four more that, with the one on the table, make a worse hand. The good cards are placed on top of the deck and the bad cards on the bottom, with the deck facing down.

Two sequences

The Magician has a small deck of cards in his hand and tells the Volunteer that he is going to play a poker game with her. Before the game, he shows her the cards, in pairs, and puts them back in the pack. He then removes the cards again in pairs, this time without showing them to the Volunteer, and asks her to choose one card from each pair, until he has five cards. When the Volunteer sees her hand, she sees that she has a sequence of nine to King. However, when the Magician shows the hand he's dealt, it turns out to also be a straight — but from 10 to Ace!

Method

The Magician has 11 cards in his hand. From top to bottom, with the deck facing down, the order should be as follows (suits are irrelevant): K Q Q 10 10 J J 9 9 A K.

The Magician should then draw pairs of cards from the top of the deck, show them to the Volunteer, and place them underneath the deck. After doing this five times, the order should be this: K K Q Q 10 10 J J 9 9 A.

Now, the Magician takes the two cards from the top of the deck again, but doesn't show them to the Volunteer, he just asks him to choose one. He deals that one (which is necessarily a King), placing it face down on the table, and places the other one underneath the deck. He repeats this four more times, so that the Volunteer is left with five cards on the table, face down. The Magician is left with six cards in his hand in the following order from top to bottom, with the deck face down: A K Q 10 J 9.

As the Volunteer looks at his hand, the Magician ostensibly places the pack in his hand on top of the rest of the deck, and deals the top five cards, which will form his hand: a sequence from 10 to Ace.

Mathematics

The initial order and the way the cards are circled allows pairs of distinct cards to be shown to the Volunteer, although the second time around the pairs that are presented to the Volunteer have cards of equal value.

Unfortunate 57

Effect

The Magician shuffles a pile of cards and asks the Volunteer to choose a suit. He announces that he will draw the first six numerical cards of that suit, which he does, placing them in a pile and emphasizing that he has drawn them in the order they came out.

Next, he shuffles the cards a bit more, asking the Volunteer to tell him when to stop. At this point, he again asks the Volunteer to choose a suit, and once again draws six number cards of the chosen suit in order. He places them in another pile, insisting that in both cases the order of the cards has not been changed.

The Magician now places the cards from the first pile, sequentially, on the table, forming a multi-digit number. He also announces a number, between 1 and 6, and asks the Volunteer to multiply the number given by the cards by that number. By placing the cards from the second pile sequentially on the table, he verifies that the number that appears is exactly the product that has just been calculated.

Method

The deck has to be prepared as follows:

We must make four piles with the following cards (face up, the order is from top to bottom), one pile for each suit: A-4-2-8-5-7.

With the stacks facing up, the top card should be the Ace. The name of the trick is a mnemonic for the order A-4-2-8-5-7.

Next, do a riffle shuffle of two of the piles, and another with the other two, now getting two larger piles.

Finally, place the face cards on top of one of the piles, in no particular order, placing the other pile on top. The cards 3, 6, 9 and 10 do not enter the trick. This ends the preparation.

When the Magician shuffles, in front of the Volunteer, he must riffle shuffle, cutting the deck in the area of the face cards: This does not change the order of the number cards. Thus, the number given by the cards drawn first is always 142857.

Next, the Magician makes a cut shuffle, holding the cards in hand, and stops when the Volunteer asks him to do so. When selecting cards of the second suit, he multiplies the value of the last card by 3 and announces the last digit of that number as the multiplier.

The order of the cards may not be the same as the other one, but the cyclical order is the same (it may not start on 1).

Mathematics

This trick is based on a curious property of the number 142857. This number is the period of the decimal representation of $1/7 = 0.(142857)$, and it turns out that numbers of the form $n/7 = n \times (1/7)$, with $1 \leqslant n \leqslant 6$, have periods with the same digits, in the same order, although each one starting with a different digit. This is naturally reflected in the multiples of the number 142857:

$$1/7 = 0.(142857) \qquad 1 \times 142857 = 142857$$
$$2/7 = 0.(285714) \qquad 2 \times 142857 = 285714$$
$$3/7 = 0.(428571) \qquad 3 \times 142857 = 428571$$
$$4/7 = 0.(571428) \qquad 4 \times 142857 = 571428$$
$$5/7 = 0.(714285) \qquad 5 \times 142857 = 714285$$
$$6/7 = 0.(857142) \qquad 6 \times 142857 = 857142$$

Returning to the trick, we recall that the number cards, of any suit, initially have the order 1-4-2-8-5-7. After the cut shuffle, the number that the number cards represent is a multiple of 142857. To find out which one, we use the property $3 \times 7 \equiv 1 \pmod{10}$ (two numbers are congruent modulo 10 when they have the same unit digit, see the appendix on modular arithmetic). So, if the last digit of $n \times 148257$ is a, then $a \equiv n \times 142847 \pmod{10}$ and

$$3 \times a \equiv 3 \times (n \times 148257) \equiv n \times (3 \times 7) = n \times 21 \equiv n \pmod{10}$$

Thus, the number n, announced by the Magician, can be obtained by multiplying the value of the last card by 3 and taking the last digit of this product.

How many cards have passed?

Effect

The Magician has a pile of 13 cards in his hand, which he cuts several times and hands, face down, to the Volunteer. He asks her to pass a certain number of cards, one by one, from bottom to top of the deck, memorizing the number. The Volunteer returns the deck to the Magician, who fans it, face down. He then states that he knows how many cards have been passed up from the deck: He turns one of the cards face up, and its value is exactly the number of cards the Volunteer has passed.

The trick can be repeated.

Method

The deck, initially, must have the cards from King to Ace (suits are indifferent), in order, with the King on top when facing down.

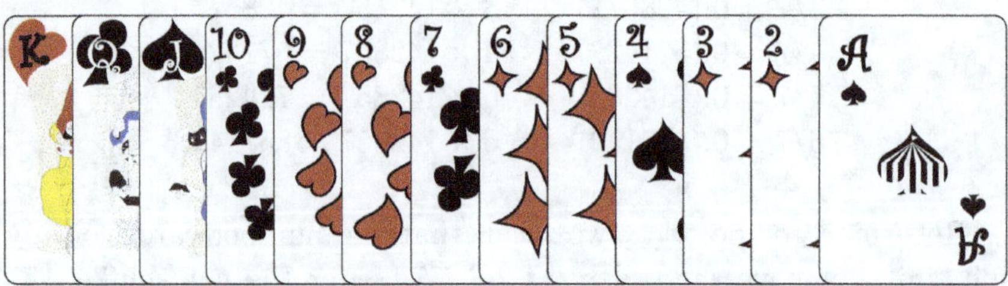

After the Magician cuts the deck, he sees what the bottom card is, before giving the deck to the Volunteer: He can, for example, do the cut shuffle in his hand, which makes it simple to see this card. Suppose it was a six. When the Volunteer returns the deck, the Magician selects the sixth card from the top.

If the Magician wants to repeat the trick, he doesn't have to look at the bottom card again: Knowing what the sixth card is, he can mentally calculate what the bottom card is by adding the value of the card with the position and subtracting 13 if necessary. In our example, if the sixth card were a two, then the bottom card in the deck would be an eight ($2 + 6 = 8$). If the card in the sixth position was a Queen, then the bottom card would be a Five: $12 + 6 - 13 = 5$.

Mathematics

All the additions and subtractions we will consider are done modulo 13 — in practice, if a result is negative, you have to add 13, and if it is greater than 13, you have to subtract 13. For those who want to know more, see the appendix on modular arithmetic.

The deck given by the Magician to the Volunteer has underneath it the card n. After passing k cards from bottom to top, the card n is exactly at position k (from the top).

Because of the way the deck is arranged, the card in position $k+t$ is the one that has value $n - t$. Therefore, to find out which card is in position n, which is the one we turn over, we first calculate t: if we want $k + t = n$ then $t = n - k$.

Therefore, the value of the card that is in position n has value $n - (n - k) = k$. To know which card is the bottom card, the principle is the same: In this case, the position is 13, and we thus have $t = 13 - k$. The card in position 13 then has value $n - (13 - k) \equiv n + k$ (mod 13). This trick is mentioned in Gardner (2014).

Magical Tic-Tac-Toe

Effect

The Magician proposes to the Volunteer a Tic-Tac-Toe game with cards. To do so, they will use cards from Ace to 9 of a suit that the Volunteer is invited to choose. After shuffling the cards, the Magician removes the cards from the deck in the order in which he finds them, emphasizing that he has not changed that order.

He now places a card on the table, face down, marking the position of the middle of the board. You ask the Volunteer to indicate the remaining eight squares with her finger, to make sure that she can locate the board, and you also ask her which square she will play next.

The Magician now places the pile of cards face down on the table. The Volunteer then takes the top card from the pile and places it on the square she announced, face up. From this point on, the Magician always plays with the cards face down, and the Volunteer plays with the cards face up, removing them from the pile one by one. The Volunteer cannot let the Magician win, so the game ends in a draw. However, after recalling that the cards have been shuffled and the suit was chosen by the Volunteer, the Magician begins to turn up the cards that are down one by one. By adding up the rows, columns, and diagonals as they are revealed, it turns out that the square is a magic square!

Method

The starting deck has to be prepared. The cards from Ace to 9 of each suit must be placed in the following order: A, 8, 2, 7, 3, 4, 5, 6, 9. If you hold the cards in your hand, as in a game, this is how the cards appear:

One way to memorize it is to start by having the cards sorted in your hand from Ace to 9, then remove the 8 and 7 and place them before and after the 2.

To prepare the deck, you make a riffle shuffle of two suits, arranged as above, and put them under the deck. Do the same to the other two, this time placing them on top. The remaining cards remain in the middle, and the deck is ready. When facing the Volunteer, the Magician makes a riffle shuffle by cutting the deck in the middle section, which does not cut through the prepared suits. Thus, the initial order is never changed, and the Magician is left with the cards of a suit in his hand in the order that is described above. Holding the cards face down in his right hand (the bottom card should be 9), he separates to his left hand the bottom three cards, placing the top card of this group of three (which is 5) face down in the middle square. If the Volunteer chooses to play on a corner square, the Magician places the two cards in his left hand on top of the others. Otherwise, he puts the two cards of his left hand under the others. In either case, he puts the deck on the table, and from now on both the Magician and the Volunteer play by drawing cards from that pile. The Volunteer sets the cards face up, and the Magician, face down.

The way the Magician plays is different, depending on whether the Volunteer's first move is in a corner or on a side. Let's describe the respective techniques. To do this, we will need to define *neighboring houses*: Two houses are neighbors if they share an edge, for example, the center is neighbor to all side houses, and each side house is neighbor to the center and two corners.

If the Volunteer's first move is in a corner, the Magician plays on a side square that is not neighboring the one the Volunteer played on. If we write the move orders on the squares, there are two possibilities for the third move:

<table>
<tr><td>2</td><td></td><td></td></tr>
<tr><td></td><td>1</td><td>3</td></tr>
<tr><td></td><td>3</td><td></td></tr>
</table>

From here on, the moves are all forced.

If the Volunteer's first move is on a side square, the Magician follows the following rule: He must always play on a square next to the one where the Volunteer has just played, and he must always attack (as is done when playing tic-tac-toe), that is, he must leave a row of two cards with the third square open, forcing the Volunteer to prevent victory.

More explicitly: After the Volunteer's first move, the Magician plays in a neighboring corner of that square (any one will do):

		3
	1	2
		3

The Volunteer's move is now forced. A possible development is as follows:

4	5	
	1	2
7	6	3

Remember that the Magician's moves are the odd ones. All the Volunteer's moves were forced, to prevent the Magician's victory.

The final result, after the cards are turned over, will always appear as shown in the figure on the next page, possibly rotated or reflected.

Mathematics

It was Martin Gardner who had the idea of creating a trick out of the magic square, and who presented this idea to Don Costello, a mathematics professor. Don Costello created various card arrangements that would achieve the effect, always involving an adjustment that depended on the Volunteer's first move. It was Dai Vernon who finally created this last arrangement, in which the adjustment can be made very discreetly, without the audience noticing. The text we have presented here is inspired by the one that appears in Gardner (2014).

The six piles

Effect

The Magician presents a deck of cards, some facing up, some facing down, and separates it into six piles in front of the Volunteer.

Now, he asks her to choose some of the piles for herself, as many as she wants (between 1 and 5). The rest are left to the Magician. From the piles chosen, the Volunteer chooses how many cards she wants to draw (between one and five) to form a magic pile, suppose

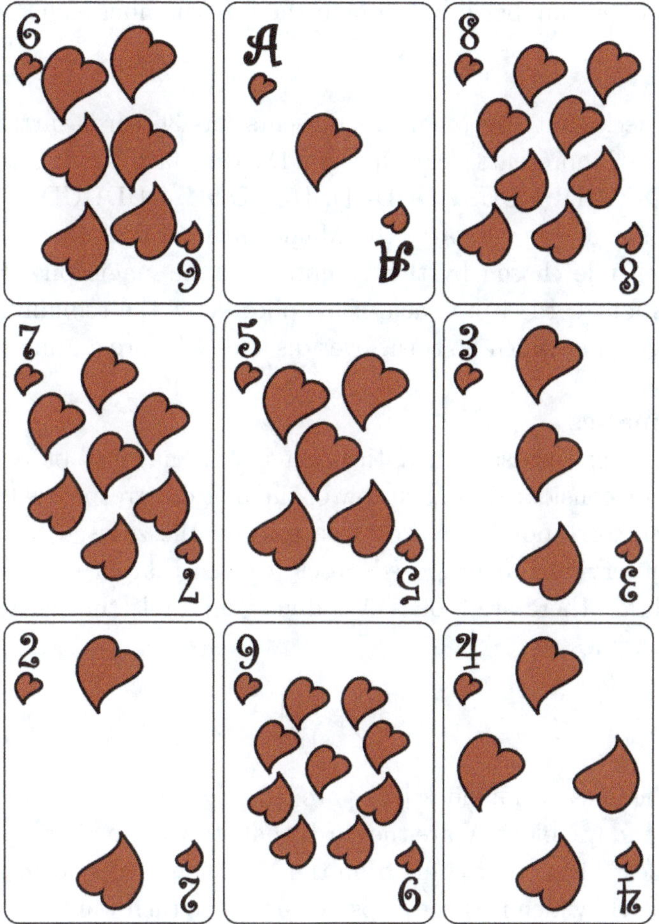

she chose four. She then deals four cards from each pile to a magic pile on the table.

For the piles rejected by the Volunteer, the choice has to be different: the Magician draws four cards from each of these piles, but discards them, keeping the rest, which are then placed in the magic pile.

After highlighting the fact that the magic pile was built with two choices of the Volunteer — the piles chosen and the number of cards — the Magician states that there was a coincidence: In the magic pile, there are as many cards face up as face down (he can even tell their number if he counted the total number as he dealt the cards).

Method

The deck that the Magician presents has 36 cards, arranged as follows — U represents a card face up, D a card face down: UDDUUD, UUDDDU, DDUUDU, UDUDUD, DUUDDU, DUDUUD.

In the example we gave, the Magician must deal the four cards from each pile chosen by the Volunteer to the magic pile. He then discards four cards from each of his piles, adds the remaining cards to a single pile, which he turns over discreetly before adding it to the magic pile.

Mathematics

This trick is based on a theorem about matrices of zeros and ones. Let's consider that each row of the board corresponds to a pile. The ones correspond to cards face up and the zeros to cards face down, observing the proposed order, we see that in each row and each column there are 3 ones. We divide the table that corresponds to the six piles as

$$\begin{bmatrix} X & Y \\ Z & W \end{bmatrix}$$

where the lines of the block $[XY]$ are the piles that the Volunteer has chosen, and the block X are the cards that stay in the Magic pile from those piles. Those that come from the Magician's piles correspond to the W block, which is turned upside down. We then want the number of ones in block X plus the number of zeros in block W to equal the number of zeros in block X plus the number of ones in block W.

Let x_0, x_1 be the number of zeros and ones, respectively, in X, and y_0, y_1, w_0, w_1 the same for blocks Y and W. Since each row and each column has 3 zeros and 3 ones, we have

$$x_1 + y_1 = x_0 + y_0 \qquad w_1 + y_1 = w_0 + y_0$$

Subtracting the second equation from the first and rearranging, we get $x_1 + w_0 = x_0 + w_1$, as wanted.

The theorem is more general — this version, created by Pedro Freitas, was enough for the performance.

Automatic counting

Effect

The Magician presents a Volunteer with a deck of cards face up, and asks her to cut it as many times as she wants. Finally, he asks her to draw a pack of cards, holding the deck in her hand. After some thought, the Magician tells how many cards are in the pack that the Volunteer took out.

Method

To do this trick, we start by preparing the deck, so that the difference between each card and the one immediately below it is 3, with the convention that, after the King, the order is cyclic and returns to A, 2, 3, ... With the deck being face up, each card should be worth three more than the card immediately below it.

To calculate the number of cards in the removed deck, you must multiply by 4 the difference between the value of the top card in the deck left on the table and the top card in the deck the Volunteer is holding. If the table card is less than the other, you can add 13 to its value to make the difference positive (but you don't have to, you can work with negative numbers).

The desired number of cards may be that value, or we may have to add or subtract a multiple of 13 to get it. It is an estimate of the size of the pack that will tell you which operation to do.

Let's present two examples. Let's suppose that on top of the table's deck is an 8, and on top of the Volunteer's deck is a 3. The difference is 5, which multiplied by 4 is 20. Then the Volunteer can have 7, 20, 33 or 46 cards in her hand.

In another example, if the top card in the table deck is a 4 and the top card in the Volunteer's deck is an 8, we add 13 to the 4, getting 17, and the difference to 8 is then 9. Multiplying by 4, we get 36, so the Volunteer can have 10, 23, 36, or 49 cards in her hand.

Mathematics

It is interesting to see that it is possible to do this sorting with all 52 cards in the deck, and that after sorting the cards in this way, the difference between the one at the top and the one at the bottom is still 3. The demonstration is based on the fact that for $1 < a < 13$

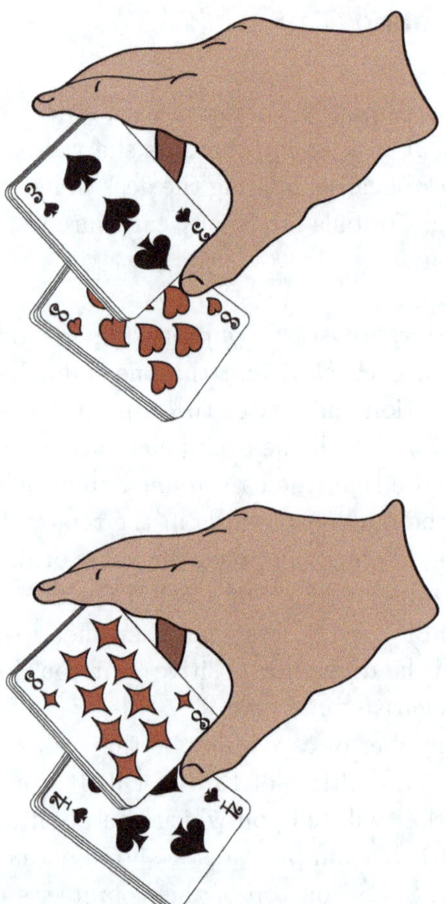

the succession a, $a + 3$, $a + 6$, $a + 9, \ldots$, $a + 36$ runs, modulo 13, through the numbers 1, 2, 3, 4, …, 13, without repetition, returning to the initial value a after making a complete tour. This allows you to cut the deck as many times as you want without changing this configuration.

Suppose then that we have the deck arranged like this, that the Volunteer has cut it and has a pack in her hand. We call this pack A, and B the one left on the table. At this point we observe the values of the two top cards of each deck, calling a the one from deck A, and b the one from deck B.

The goal is to find out the number of cards in pack A, knowing the values a and b. Let's call this number n.

Actually, what we will know is how many cards are in pack A modulo 13, that is, the value we will get will be of the type $n + 13t$ with $t \in \mathbb{Z}$, as we explained. It is common sense that tells us what value we have to use for t.

The number $a - b$ is exactly the triple of this number (mod 13) because from each card to the next this difference increases by 3 units. Thus, $a - b \equiv 3n$ (mod 13) and therefore $b - a = -(a - b) \equiv -3n$ (mod 13). Now, multiplying the difference by 4, we get

$$4(b - a) \equiv 4 \times (-3n) = -12n \equiv n \quad (\text{mod } 13)$$

since $-12 \equiv 1$ (mod 13), which justifies the algorithm of multiplying the difference by 4 to get the number of cards in deck A, modulo 13.

Variant: Those who find arithmetic in modulo 13 difficult can use a deck of cards from 1 to 10 of all suits, for a total of 40 cards. The deck is prepared in the same way, with a difference of three from one card to the next, and to get the number of cards from the top deck, multiply by 3, and the result obtained gives the desired number modulo 10, for example, if the number obtained is 15, the deck will have 5, 15, 25 or 35 cards. In this variant, if the card in the table deck is of small value, 10 must be added before calculating the difference.

Sequeira divination

Effect

The Magician holds a deck of cards in his hand, which he opens in a fan, with the faces of the cards facing down, and asks the Volunteer to draw a card and look at it carefully, without showing it to the Magician. The Magician guesses the card.

Method

The deck must be prepared, with the values in arithmetic progression with common difference 4. With the deck face up, the value of a card is that of the bottom card plus four units, going around when it reaches the King (after the King comes the 4). In the original order, the Queen is worth 11, the Jack 12 and the King 13.

The suits are ordered in the order CHaSeD. Thus, from bottom to top, and with the deck facing up, the sequence is

A♣5♥9♠R♦4♣8♥V♠3♦7♣D♥2♠6♦10♣A♥5♠9♦R♣4♥8♠V♦3♣7♥D♠2♦6♣10♥
A♠5♦9♣R♥4♠8♦V♣3♥7♠D♦2♣6♥10♠A♦5♣9♥R♠4♦8♣V♥3♠7♦D♣2♥6♠10♦

When the Volunteer draws a card, the Magician subtly cuts the deck at the place of the chosen card, completing the cut, and looks discreetly at the card underneath. Now, he simply adds 4 to the value of the card and thinks of the suit that comes next, in order, to know which card the Volunteer is.

Mathematics

This order involves modular arithmetic, modulo 13, and the fact that an arithmetic progression with common difference 4 goes through all the modulo 13 classes before returning to the initial class (see appendix on "prepared decks").

This effect is mentioned in de Sequeira (1612).

Sequeira spelling

Effect

The Magician, after shuffling the cards, asks the Volunteer to choose one, separating it from the deck and placing it on the table, face down. He then states that he is unable to guess which card it is, but that the deck will be able to do so. He then says the words "WHAT SUIT", drawing one card per letter from the deck (reversing the order), into a pile of eight cards, which he keeps face down. He does the same with the word "VALUE", forming a pile of five cards. Then he makes a third pile, repeating the process with the words "COMPANION CARD".

He then turns over the top cards from each of the piles. The one in the first pile will be of the same suit as the chosen card, and the one in the second pile will be of the same value. The one in the third pile will be of the same color and value as the chosen card. When you turn the chosen card over, everything checks out!

Method

You must have the deck sorted as in the previous effect, or with a progression order, the initial shuffle must be a cut shuffle. When the

Volunteer chooses his card, you must cut the deck at that point and complete the cut. From here on the trick is automatic.

Mathematics

Two cards at a distance of 8 from each other are of the same suit, and if they are at a distance of $13 = 8 + 5$ they have the same value.

This effect appears in one of the videos of Brushwood (n.d.).

Two cards indicate another

Effect

The Magician fans the deck in his hand, face down, and asks the Volunteer to choose two cards, next to each other, and show them. Suppose the Volunteer has selected the six of spades and the 10 of diamonds. The Magician says that these cards can indicate another, one of them indicating the suit and the other the value. The two cards drawn, for example, can indicate the 10 of spades or the six of diamonds. The Volunteer says which one he wants, and the Magician makes the chosen card appear.

Method

The deck must be prepared with the sequence of the previous tricks (or Stebbins', see appendix). The Magician must cut the deck at the place of the chosen cards and put it back together again, completing the cut. The possible cards chosen are now in place 12, either from the bottom or from the top. In the example given, and keeping the deck face down, the seven of diamonds is the 12th card from the top, and the 10 of clubs is the 12th from the bottom.

To find out where each card is, the following rule can be used: If it is the lowest card that gives the value, start at the top, if it is the highest, start at the bottom (always with the deck face down). The term "lowest card" is used in the sense of Sequeira's order, for example, if a Queen and a three come out, the Queen will be the lowest card.

The Magician can have a written prediction, such as "Your card is at position 12", for example, or he can deal cards to the table, without seeing them, and stop at the 12th, turning it over after placing it on the table.

The way the cards are arranged in this sequence generates several regularities: The suits are repeated every four cards and the values every 13. Thus, the chosen card is 13 cards away from the card that gives the value and 12 from the card that gives the suit.

This effect appeared described on the YouTube channel Mismag822 (n.d.), using Stebbins' order.

Where is the card?

Effect

After shuffling the cards, the Magician asks a Volunteer for the name of a card. Putting the deck down on the table and giving it some thought, he says in what position the requested card is (from the top of the deck).

Method

The deck must be prepared as in previous tricks, but this time with an arithmetic progression with common difference 3, that is, the value of each card is three units higher than the card below it, with the deck facing up. The suits are arranged in the order CHaSeD. From bottom to top, and face up, the order is as follows:

A♣4♥7♠10♦R♣3♥6♠9♦D♣2♥5♠8♦V♣
A♥4♠7♦10♣R♥3♠6♦9♣D♥2♠5♦8♣V♥
A♠4♦7♣10♥R♠3♦6♣9♥D♠2♦5♣8♥V♠
A♦4♣7♥10♠R♦3♣6♥9♠D♦2♣5♥8♠V♦

After the Volunteer says the name of the card, the Magician looks for the card of the same suit as close to the bottom of the deck as possible (face down) before placing it on the table. He then subtracts from the value of the card from the deck, the value of the card requested, and multiplies the result by 4 — if the value of the card in the deck is smaller, he adds 13 to it before counting. From this value, he subtracts the number of cards that are under the observed card in the deck, and that is the desired number.

For example, if the card requested is the deuce of clubs and the closest card of clubs to the bottom of the deck is the seven, which is behind two cards, then the Magician calculates $7 - 2 = 5$, $4 \times 5 = 20$, $20 - 2 = 18$, then says that the deuce of clubs is the 18th card from the top of the deck.

If the card requested is the Jack of clubs, with the same deck, the Magician does the following math: $7 + 13 = 20$, $20 - 11 = 9$, $4 \times 9 = 36$, $36 - 2 = 34$, and announces that the card is in the 34th position.

Mathematics

What is remarkable about this trick is that, unlike in "Automatic counting", the result is exact and no multiple of 13 needs to be added or subtracted.

For the explanation, let's first look at a special case: Suppose the card on the bottom of the deck is of the same suit as the requested card. Let's call a the requested card, b the card on the bottom of the deck, and let's further suppose that $a > b$. In this case, the rule says that the position of card a (from the top of the deck) is given by

$$4 \times ((13 + b) - a) = 52 + 4 \times (b - a)$$

From the way the cards are arranged, we see that it takes four cards to go from one card to the next card of the same suit. Because of this, the number of cards below a is $4 \times (a - b)$. Thus, the position of the card a is $52 - 4 \times (a - b)$, which is the number given by the rule.

If $a < b$ the rule says that the position is given simply by $4 \times (b - a)$. To confirm, we pass card b provisionally to the top of the deck, and using the previous rule, we find that under that card to card a (including card a, but excluding b) there are exactly $4 \times (b - a)$ cards. If we put card b back on the bottom of the deck, this number becomes the position of card a.

Finally, if card b is not the bottom card in the deck, we can move the cards underneath that one to the top of the deck, do the count, and put them back underneath at the end, which requires subtracting that number to get the position of card a.

This effect is described in Si Stebbins' famous late 19th century pamphlet Stebbins (2019), which presents various effects that can be done with the deck prepared in this way.

Three in one

Effect

The Magician throws the deck into the audience, held together by rubber bands, or asks for three Volunteers. After cutting the deck as many times as desired, the Volunteers each take their card from the top of the deck, in order. The Magician then announces that he is going to try to see what the cards are (if the Volunteers are in the back of the room, the effect is even better). Simulating difficulty in seeing, he asks the first one to tell him the value of the card. Since the difficulties are not over, he asks the second one to tell him the suit of his card. From this point on, the Magician tells what the three cards are without any further information.

Method

As with the previous effect, you simply sort the deck by putting the suits in CHaSeD order and the card values in a way that the Magician can remember, such as the order of the progression or Eight Kings. With the value of the first card, the Magician knows all the values, with the suit, he knows all the suits.

Mathematics

There is not much to say: Knowing the rule of both sequences (suits and values), knowing one term you can easily know what the neighboring terms are.

Joining by suits

Effect

The effect is similar to the previous one. The Magician throws the deck of cards, held together with a rubber band, into the audience and asks that it be passed from hand to hand until it reaches as far away as possible. He then asks the Volunteer currently holding the deck to cut it and give it to the person to his right. The Magician asks them to do the same, and pass it to the next person. This is repeated until the fifth person is reached. The Magician then asks that Volunteer to take the top card from the deck and pass it to the person to his left. That Volunteer must also take the top card from the deck, and so on, until all five Volunteers have a card in their hand.

The Magician announces that he is going to guess the five cards. Simulating difficulty, he asks people their names, writing them down in a notebook. He then asks people to group by suits — those with cards of the same suit should be in the same group — but without being told which suits they are. After a little more writing, the Magician guesses the five cards correctly.

Method

As with the previous trick, the deck must be prepared. The order of the cards must be as shown at the end of the text.

After the Volunteers have chosen the cards, the Magician writes down the names of the Volunteers in order, writing the numbers 1–5 next to the names. In the notebook, he must have the table below where he sees which cards correspond to the pattern formed by the grouping of the Volunteers. This must be done without the audience realizing that the Magician is reading a list. The table has been organized to try to make this search easier, for example, in the first column you see the partitions that have a set of two cards. In general, we have ordered the partitions according to the size of the largest subset.

We used vertical dashes to define the subsets, for example, $13\,|\,24\,|\,5$ corresponds to the partition $\{1,3\}$, $\{2,4\}$, $\{5\}$. The cards are arranged in the right column of the table in the order in which they were drawn from the deck, to facilitate the Magician's final answer. For example, the following row of the table means $1 = 10\heartsuit$, $2 = 9\clubsuit$, $3 = 2\heartsuit$ $4 = 5\clubsuit$ and $5 = 8\spadesuit$. The Magician can write in the notebook each Volunteer's card next to their name so the final answer is quicker.

Mathematics

This trick is first mentioned in Diaconis and Graham (2015) and was inspired by a mathematical concept called *Bell numbers*. Bell numbers, B_n, tell you how many partitions there are of a set with n elements, that is, how many ways you can decompose it into pairwise disjoint subsets. For example, the set $\{1,2,3\}$ has five partitions:

$$\{1\}\{2\}\{3\} \qquad \{1,2\}\{3\} \qquad \{1,3\}\{2\} \qquad \{2,3\}\{1\} \qquad \{1,2,3\}$$

Thus, $B_3 = 5$. The authors of the book then noted that $B_5 = 52$, which is exactly the number of cards in a deck, and in view of such a coincidence, they invented this effect. The fundamental idea was to define the five-card partition of the deck by putting cards of the same suit together. It was then necessary to find an order in the 52 cards with the property that each five consecutive cards would give rise to a distinct partition. Since with four suits it would be impossible to get the partition into five sets (each one with only one card), it was necessary to add the Joker. After some work, the authors, in collaboration with Fan Chung, proposed the order of suits that we present here, which has the desired property. We proposed the Eight Kings order within each suit and constructed the respective table.

In this case, it was the mathematical coincidence that prompted the invention of the magic effect.

The order is from bottom to top, with the deck face up; the Jack of spades is replaced by the Joker, which we represent by *J:

8♦K♦3♦10♦2♦8♣8♥K♥3♥K♣3♣7♦9♦
10♣2♣7♣10♥9♣2♥5♣8♠7♥9♥K♠5♦3♠
10♠Q♦2♠7♠5♥9♠4♦A♦Q♣Q♥5♠Q♠4♣
4♥4♠A♥6♦6♥A♠A♣J♥6♠*J 6♣J♦J♣

Sequences

1|2|3|4|5	J♥6♠*J 6♣J♦	123|4|5	3♦10♦2♦8♣8♥
12|3|4|5	4♦A♦Q♣Q♥5♠	124|3|5	2♠7♠5♥9♠4♦
13|2|4|5	A♥6♦6♥A♠A♣	134|2|5	10♠Q♦2♠7♠5♥
14|2|3|5	A♠6♠A♣J♥*J	234|1|5	9♦10♣2♣7♣10♥
15|2|3|4	A♣J♥6♠*J 6♣	125|3|4	5♠Q♠4♣4♥4♠
23|1|4|5	9♠4♦A♦Q♣Q♥	135|2|4	4♥4♠A♥6♦6♥
24|1|3|5	4♣4♥4♠A♥6♦	235|1|4	Q♦2♠7♠5♥9♠
25|1|3|4	6♦6♥A♠A♣J♥	145|2|3	2♥5♣8♠7♥9♥
34|1|2|5	5♥9♠4♦A♦Q♣	245|1|3	9♥K♠5♦3♠10♠
35|1|2|4	6♠*J 6♣J♦J♣	345|1|2	2♦8♣8♥K♥3♥
45|1|2|3	A♦Q♣Q♥5♠Q♠	123|45	8♥K♥3♥K♣3♣
12|34|5	K♥3♥K♣3♣7♦	124|35	2♣7♣10♥9♣2♥
12|35|4	7♥9♥K♠5♦3♠	134|25	K♠5♦3♠10♠Q♦
12|45|3	10♦2♦8♣8♥K♥	234|15	8♣8♥K♥3♥K♣

13|24|5	10♥9♣2♥5♣8♠	125|34	K♣3♣7♦9♦10♣
13|25|4	9♣2♥5♣8♠7♥	135|24	7♣10♥9♣2♥5♣
13|45|2	7♠5♥9♠4♦A♦	235|14	5♦3♠10♠Q♦2♠
14|23|5	8♠7♥9♥K♠5♦	145|23	3♣7♦9♦10♣2♣
14|25|3	6♥A♠A♣J♥6♠	245|13	6♣J♦J♣8♦K♦
14|35|2	Q♠4♣4♥4♠A♥	345|12	7♦9♦10♣2♣7♣
15|23|4	Q♥5♠Q♠4♣4♥	1234|5	K♦3♦10♦2♦8♣
15|24|3	4♠A♥6♦6♥A♠	1235|4	10♣2♣7♣10♥9♣
15|34|2	Q♣Q♥5♠Q♠4♣	1245|3	3♠10♠Q♦2♠7♠
23|45|1	3♥K♣3♣7♦9♦	1345|2	J♦J♣8♦K♦3♦
24|35|1	*J 6♣J♦J♣8♦	2345|1	J♣8♦K♦3♦10♦
25|34|1	5♣8♠7♥9♥K♠	12345	8♦K♦3♦10♦2♦

Where are the red ones?

Effect

As with the previous two effects, the Magician throws the deck of
cards, attached by a rubber band, into the audience, so that it is as
far away from you as possible to make sure you don't see any cards.
He asks five Volunteers, standing side by side, to cut the deck. After
the five cuts, each chooses a card, in turn, from the top of the deck.

The Magician announces that he is going to guess the five cards.
Simulating difficulty, he asks whoever has a red card to raise their
hand. He can then take a small notebook and write some things, or
do math with his fingers, always saying that the task is very difficult
(he can also ask more questions, such as people's names, or where
they live). After this, he guesses the five cards correctly.

Method

The deck has only 32 cards (or 31 in another version), from Ace
to eight of each suit, and must be arranged as described at the end
of the text.

Associating 0 to black and 1 to red, the chart that we also present
at the end of the text matches the possible color sequences with the
value of the cards.

The Magician can then have this chart in his notebook, look
quickly, and, given the pattern of zeros and ones, find the sequence.

It's also a good idea to write down the names of the cards and then quickly say them. It is also important not to spend too much time looking at the board: The Magician must not make it seem as if he is looking at a table. Note that in the first part of the table, all the patterns begin with 00, in the second part with 01, in the third part with 10, and in the fourth part with 11. Within each part of the table, the first four rows have a 0 in the third position and the second rows have a 1. This can help speed up the search.

There is also a way to do the trick without any table, but for that the Magician has to know the binary system.

Given five numbers, we will agree that the first two define the suit and the last three define the value of the card. For suits, the convention is: $00 =$ clubs, $01 =$ spades, $10 =$ diamonds, $11 =$ hearts.

So, we see that the first number says whether the suit is black (0) or red (1). Considering the order of suits in bridge, we can regard clubs and diamonds as minor suits and spades and hearts as major suits. With this convention, the second number says whether the suit is lower (0) or higher (1).

The next three digits are the value of the card in binary. For example, 110 represents $4 + 2 = 6$. The only exception is the eight, which is encoded by 000.

So, we see that each pattern of zeros and ones in the above table encodes the first card in the pattern! Therefore, when the Magician sees where the red cards are, he immediately knows the card of the person on the left. How does he now know the others?

Again, the order in the deck helps. Let's consider the sum modulo 2 in the set $\{0, 1\}$:

$$0 + 0 = 0 \qquad 0 + 1 = 1 \qquad 1 + 0 = 1 \qquad 1 + 1 = 0$$

Given five numbers, *abcde*, representing the colors of five consecutive cards in the deck, the color of the next card is given by $a + c$ modulo 2. So, if the pattern is 10010 (two of diamonds), the next digit is $1 + 0 = 1$, and the next card is given as 00101 (five of clubs). The sequence then becomes 100101. The next number is given by the sum of the marked numbers: **1**0**0**101, is $0 + 1 = 1$. In general, each following number is given by the sum of the third and fifth numbers from the end. Continuing this algorithm to the end, we get 100101100, where each sequence of five consecutive digits gives a card:

$$10010 = 2\diamond \quad 00101 = 5\clubsuit \quad 01011 = 3\spadesuit \quad 10110 = 6\diamond \quad 01100 = 4\spadesuit$$

So, by doing some math, the Magician can find out the remaining four cards. However, to use this rule you must use a deck of 31 cards, removing the eight of clubs (coded by 00000)! To find out why, see the paragraph on the mathematics of this trick.

These computations can be done in the notebook or, with some practice, using the hands. The Magician can use his fingers to code the zeros and ones, for example, he can put his hands on the table, or on his leg, representing a 0 by a bent finger and a 1 by a stretched finger. Knowing the colors of the Volunteers' cards, he can place the fingers of his left hand appropriately, according to that pattern. From there, he can successively calculate the positions of the fingers of the right hand (nine are enough for the whole sequence), or go on calculating card by card, using only the left hand. With all fingers in the right position, the Magician can then read the values and suits of the cards. Although some spectators may perceive this as coding, the process is quite mysterious and surprising!

Mathematics

As we have seen, the order defined in this deck is truly ingenious! It combines two rather delicate mathematical concepts: de Bruijn sequences and the so-called linear shift registers, used in computer science. A sequence of zeros and ones is a de Bruijn sequence with a window of size k if each sequence of k digits appears once and only once in the sequence. The sequence we present is a sequence of length 32 and window size 5.

Besides having this property, this sequence is also generated as a linear shift register, that is, each digit is obtained through a simple rule, starting from the previous digits. At this point, we invite the reader to write down five digits (zeros and ones), and generate the sequence using the rule we have described. For example, if you start with 11010, the next digit is $1 + 0 = 1$ (the sum of the first and the third). The sequence becomes 110101. The next digit is $1 + 1 = 0$ (the sum of the third and fifth digits from the end).

When you get to 31 digits, you will see that the sequence you get is the one we presented and that any five-digit sequence appears in de Bruijn's sequence one and only once, except for 00000.

We found this trick, with this implementation and explanation, in Diaconis and Graham (2015).

Order is bottom to top, with the deck facing up:

8♣, A♣, 2♣, 4♣, A♠, 2♦, 5♣, 3♠, 6♦, 4♠, A♥, 3♦, 7♣, 7♠, 7♥, 6♥, 4♥, 8♥, A♦, 3♣, 6♣, 5♠, 3♥, 7♦, 6♠, 5♥, 2♥, 5♦, 2♠, 4♦, 8♠, 8♦

Sequences

00000	8♣A♣2♣4♣A♠		01000	8♠8♦8♣A♣2♣
00001	A♣2♣4♣A♠2♦		01001	A♠2♦5♣3♠6♦
00010	2♣4♣A♠2♦5♣		01010	2♠4♦8♠8♦8♣
00011	3♣6♣5♠3♥7♦		01011	3♠6♦4♠A♥3♦
00100	4♣A♠2♦5♣3♠		01100	4♠A♥3♦7♣7♠
00101	5♣3♠6♦4♠A♥		01101	5♠3♥7♦6♠5♥
00110	6♣5♠3♥7♦6♠		01110	6♠5♥2♥5♦2♠
00111	7♣7♠7♥6♥4♥		01111	7♠7♥6♥4♥8♥
10000	8♦8♣A♣2♣4♣		11000	8♥A♦3♣6♣5♠
10001	A♦3♣6♣5♠3♥		11001	A♥3♦7♣7♠7♥
10010	2♦5♣3♠6♦4♠		11010	2♥5♦2♠4♦8♠
10011	3♦7♣7♠7♥6♥		11011	3♥7♦6♠5♥2♥
10100	4♦8♠8♦8♣A♣		11100	4♥8♥A♦3♣6♣
10101	5♦2♠4♦8♠8♦		11101	5♥2♥5♦2♠4♦
10110	6♦4♠A♥3♦7♣		11110	6♥4♥8♥A♦3♣
10111	7♦6♠5♥2♥5♦		11111	7♥6♥4♥8♥A♦

Short Biographies

Alex Elmsley

Alex Elmsley was born on March 2, 1928, in Scotland, and died at the age of 76 on January 6, 2006. He studied physics and mathematics at Cambridge University, was a member of the prestigious *The Pentacle Club*, was a member of the University's Theatre Club, and worked for a British computer company. He was a computer expert, but magic was one of his great passions. He became interested in this activity very early, around 1946, when he received a magic set as a present during a hospital stay, and from then on he never stopped, reaching his peak years between 1949 and 1959. He did tricks with almost everything from cards, coins, ropes, billiard balls, cigarettes, newspapers, computers, and so on. He invented countless tricks that revolutionized magic, such as: "Ghost count" or "Elmsley count" as it was later called, "Between your palms", "Point of departure", and "Diamond cut diamond". He has a few publications: *Low Cunning* (1957 and 1959) and *Cardwork* (1975). There are also two volumes written by Stephen Minch that compile, in a way, all of Elmsley's works: *Collected Works of Alex Elmsley*.

Bob Hummer

Bob Hummer is considered by many to be a genius. He invented countless mathematical magic tricks despite having no specific training in mathematics. He was born on January 25, 1906 and died in April 1981. Bob Hummer always lived on the poverty line, when he needed money he was busking in bars in Chicago, the city where he lived, where he received tips from the people he performed for. Bob Hummer worked for a short period of time as an assistant to card magician Paul LePaul. LePaul called him on stage as a "Volunteer" and he naturally had everyone laughing. Hummer, in addition to inventing mathematical tricks, could juggle spectacularly with the cards. His most famous trick is called "Mathematical three card Monte", and it has countless variations that have inspired many other famous magicians. Same reason as before. Hummer left us many tricks involving face-up and face-down cards in his famous 1942 book *Face Up Face Down Mysteries*.

Dai Vernon

Dai Vernon was born in Ottawa, Canada on June 11, 1894, and died on August 21, 1992. Dai Vernon is the stage name of David Frederick Wingfield Verner and is considered one of the greatest magicians of the 20th century, known as "The Professor". He developed and improved some magical effects and became known as the legend of the sleight of hand. Although many of the tricks Vernon invented were mathematical, he was not very comfortable with this subject in school. He was an expert in "close-up" magic with cards, as the very funny episode with the magician Harry Houdini shows. Houdini claimed that he could discover the secret of any trick that was performed in front of him and repeated a maximum of three times. Vernon accepted the challenge and did a trick called "The ambitious card" and repeated it eight times without Houdini realizing how it was done. The first trick he learned to do was taught to him by his father, when he was only six years old, and he then fell in love with magic.

Lennart Green

Lennart Green was born in Sweden in 1941. He is world famous for his seemingly chaotic and erratic style. In 1991, he won the title of Fédération Internationale des Sociétés Magiques (FISM) World

Champion, in a way that is worth noting. In the 1988 FISM contest, Green had been disqualified on suspicion that he had used illegal help in shuffling the cards in his trick. Lennart competed again in the next contest in 1991, but insisted that the judges shuffle the cards themselves. He repeated the 1988 act and won first place.

Lennart Green is a frequent presence at the most important gatherings of magic and recreational mathematics around the world. In particular, he often performs and lectures at the *Gathering for Gardner*, both in the USA and in Portugal.

Martin Gardner

Martin Gardner was born on October 14, 1914 in Tulsa, Oklahoma. His mother was an elementary school teacher and his father a geology Ph.D. who was self-employed in oil exploration. Martin learned to read for himself from the books about Oz by Frank Baum that his mother read to him (he would later also write a book about Oz).

At school, Martin was mainly influenced by two teachers. M. E. Hurst, a physics teacher, motivated him so much that he planned to study science and mathematics later at university. His interest in magic was motivated by the card tricks his father showed him and by contact with a local magician.

Since Caltech only accepted students with general undergraduate degrees, Gardner, who intended to study physics, opted for a degree in philosophy at the University of Chicago. Here, he was a student of R. Carnap, whom he greatly admired and who transmitted to him a great interest in the philosophy of science. Another of his heroes (to use his words) was Bertrand Russell, who had a debate in Chicago with Carnap, which Gardner attended, where they disagreed about the objective existence of their wives.

In college he ended up not taking any math courses, which he later acknowledged was helpful to him! As he said in an interview in 2005, "If you write about popularizing mathematics, you shouldn't know too much," going through the trouble of understanding the subjects helped him, through very clear writing, to expose them to the layman.

In December 1956, a paper on Hexaflexagons (invented by some Princeton students, including R. Feynman) appeared in the pages of *Scientific American*, which was a success. In January 1957, Martin Gardner started his regular column "Mathematical Games" which ended in 1981.

Over these 25 years, Martin Gardner has become the greatest exponent of mathematical outreach. Some results and some mathematicians were exposed by Gardner for the first time. This Mathematical Games column has had unparalleled success, both among mathematicians and laymen. As Graham said, "He turned thousands of children into mathematicians and thousands of mathematicians into children."

One of his great virtues was to relate mathematics to many other areas, in a natural, but never superficial way. His passion for recreational mathematics led him to advocate its use in the classroom as a method to captivate students. In one of his interviews, he tells how, at a very young age, he was surprised by his math teacher when he was investigating the game of tic-tac-toe. He was scolded, "Here in class we only deal with mathematical matters." His opinion, many years later, was still that the game of tic-tac-toe is a great didactic tool, that it allows teaching about various curricular subjects in an enjoyable way. Moreover, he stressed, "Who doesn't know this game?"

Interested in magic from an early age, he has published many articles and books on the subject, and has always remained an active member of the magic community.

His knowledge in this area helped him in the task, to which he dedicated himself diligently for many years, of exposing impostors. The so-called "psychics" often resorted to tricks well known to Gardner and others. One successful charlatan, Geller, used an artifice common among magicians to "bend spoons". In this vein, he collaborated with the *Skeptical Inquirer*, publishing many articles there.

Recognition for Martin Gardner is general, but from mathematicians and magicians, he receives special affection.

A gathering on Recreational Mathematics and Magic has been held regularly in his honor in Atlanta, USA, since 1993. This is the

Gathering for Gardner (G4G) which recently held its 15th edition (G4G15) in March 2024.

A similar European meeting is held every odd-numbered years in Portugal, the *Recreational Mathematics Colloquia — G4G Europe*.[1]

Persi Diaconis

Persi Warren Diaconis was born in New York City in 1945 into a family of musicians. He studied violin at the famous Juilliard School in addition to attending the local high school. At the age of 14, his passion for card magic spoke louder, and he joined Dai Vernon, the famous Canadian magician, on a tour of the USA.

Later, already a solitary card professional, he felt the need to learn mathematics in order to read a book on probabilities that seemed important to him. Fortune, which always accompanies the daring, did not hold Diaconis back. George Washington High School gave him his high school diploma, although he had been absent for the last few years, perhaps because he had been a good student before he became a bouncer. So, he enrolled and graduated from the City College of New York in 1971 with a degree in mathematics.

Martin Gardner, whom he had met in magic circles, recommended him to Frederick Mosteller, a professor at Harvard, who accepted Persi Diaconis into his doctoral program. And so began the career of the mathematician, now at Stanford.

In addition to many important scientific works, Persi Diaconis has devoted himself to recreational mathematics, particularly that related to card tricks.

---

[1] https://ludicum.org/rmc/

Appendix

Ways of shuffling

False

Draw k cards, one by one, to the table or to your hand, reversing the order, and put them under the deck. After doing this four times, the order of the cards stays the same, as long as k is at least half the number of cards.

Let's show an example with nine cards (arranged vertically), where we pass seven cards from the top to the bottom of the deck. We will number the cards from 1 to 9:

1	8	2	9	1
2	9	1	8	2
3	7	3	7	3
4	6	4	6	4
5	5	5	5	5
6	4	6	4	6
7	3	7	3	7
8	2	9	1	8
9	1	8	2	9

The horizontal lines mark, in the corresponding columns, the change in the order of the cards — increasing or decreasing. The small groups that appear at the top and bottom of the deck have $2 = 9 - 7$ cards; in general, in a deck with n cards where we pass k cards down, the groups have $n - k$ cards.

Dovetail aka riffle

Separate the deck into two packs, holding one in each hand, and fan each into the other. If both packs contain 26 cards, and if you can mix the cards perfectly, i.e., alternating one from each pack, the riffle shuffle is said to be perfect. If a perfect riffle shuffle is performed eight times on a deck of 52 cards, always keeping the top and bottom cards in these positions, the order of the cards is restored. The way cards are shuffled is the same as in the Faro shuffle.

Reverse aka in-and-out

The reverse shuffle, where you draw one card to one pile and another to another is called the reverse riffle shuffle, or in-and-out and appears in the tricks "31 cards", "The footsteps of the thief", and "The whispering Joker". To perform it, you hold the deck in your left hand, draw the top card to the front (out), the next to the back (in), and so on, holding both the in and out packs with your right hand and adding the cards alternately under the packs. This process generates two piles of cards in the right hand:

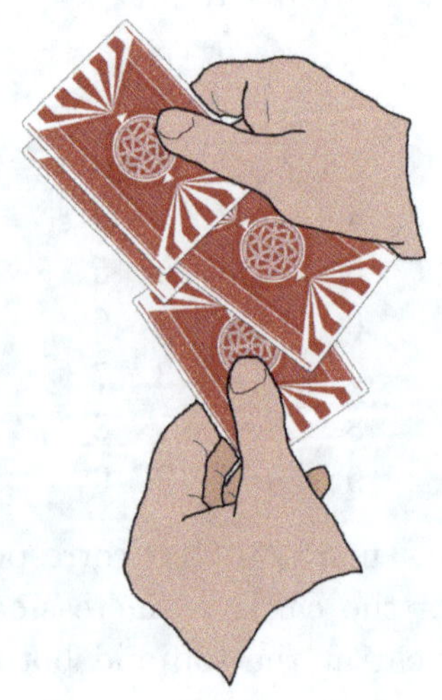

Rosetta

Separate the deck into two piles, placed on the table, and rotate them so as to make two rosettes, which are then joined together. The effect is the same as the dovetail or riffle shuffle.

Cut

Take the cards in the left hand and take, with the right hand, a pile of cards from the bottom of the deck and pass it to the top. The effect is the same as that of a cut.

Monge

The name comes from the mathematician Gaspard Monge, known for his work in geometry. It consists in picking up the deck of cards with the left hand and taking cards from the top, placing them successively on top and underneath the deck that accumulates in the right hand. We will present an example with six cards:

1											6
2	2							4		4	4
3	3		3	2		2		2		2	2
4	4	1	4	1	4	1		1		1	1
5	5		5		5	3	5	3		3	3
6	6		6		6		6		6	5	5

It is the inverse of the Klondike or milk shuffle: If you do one after the other, the cards stay in the same position. It also has the property that, when performed on a deck of $2n$ cards that are initially paired (for example, two black fives followed by two red Queens, followed by two red nines, and so on), it places the deck in mirror image (see what happens in the example above, thinking that the pairs are 1–2, 3–4 and 5–6). If it is run again, each card will be at distance n from its pair, that is, if the deck has for example 20 cards, each card is

10 cards away from its pair. This property is used in the "Daisy's socks" trick.

Milk aka Klondike

Simultaneously draw a pair of cards from the top and bottom of the deck and lay them down on the table, repeating until the deck runs out. Let's again take an example with six cards:

1					3
2	2				4
3	3		3	2	2
4	4		4	5	5
5	5	1		1	1
6		6		6	6

This is the inverse of the Monge shuffle: If done one after the other, the deck stays the same. As such, if it is performed twice on a deck of $2n$ pairs of cards, where each card is n cards away from its pair, the result is a deck with the cards paired, that is, each one next to its pair.

Australian aka down-and-under

You hold the deck in your hand and alternately draw one card to the table and one card to the bottom of the deck. Again, we present an example with 6 cards, the left column is the deck, held in the hand, and the right column represents the cards on the table:

1																		
2	2		3															6
3	3		4		4		5							2		2		2
4	4		5		5		6		6	5	2	5		5		5		5
5	5		6		6	3	2	3	2	3	4	3	4		6			3
6	6	1	2	1	2	1	4	1	4	1	6	1	6	1	4	1	4	1

When this shuffle is done until only one card is left in the hand, in a deck with n cards, this card is the one in position p given by twice the difference between n and the highest power of 2 less than n. In the case shown, with six cards, the highest power of 2 less than 6 is 4, and therefore the card left in the hand is the one in position $2 \times (6 - 4) = 4$.

It is not difficult to verify this property if n is a power of 2. In this case, after we make a complete circuit, circulating all the cards, we throw away (to the table) half the cards, keeping in the hand a number of cards that is a power of 2. Moreover, the card that was in the last position of the deck stays in the same position. After another circuit, the same phenomenon happens. After a certain number of completed circuits, we are left with two cards in our hand, and the second is the one that was originally underneath the deck. After you throw the first one on the table, that's the one that stays in your hand: The card that was originally underneath the deck in position n.

The number given by the formula is also n. If $n = 2^t$, then the greatest power of 2 less than n is 2^{t-1} and twice the difference therefore gives $2 \times (2^t - 2^{t-1}) = 2^t = n$.

In the general case, where n does not have to be a power of 2, we have to see how many cards we have to circulate until the number of cards in the hand is a power of 2 — in which case we know what happens. If $n = 2^t + k$, where $k < 2^t$, we have to draw k cards from the deck until the deck is 2^t cards. This is achieved by circulating $2k$ cards (k to the table and k down). The card that is in position $2k$ then goes to the bottom of the deck, and that will be the card that is left at the end. This justifies the formula: The position of the remaining card is given by twice the difference between n and the highest power of 2 less than n.

Hummer

Keeping the deck face down, cut and turn the top two cards — the reason for the abbreviation CATO (cut and turn two). This shuffle, when done in decks with an even number of cards, keeps the number of face-up cards in the even positions equal to the number of face-up cards in the odd positions.

Prepared decks

Suits

A possible sequence of suits in a prepared deck is coded by the word CHaSeD, which refers to the English suit names: clubs, hearts, spades, and diamonds. The reverse order is given by the word SHoCkeD. You can also use the suit order in Bridge: clubs, diamonds, hearts, and spades (from lowest to highest).

Progression

The values of the cards are placed in arithmetic progression with common difference, that is, each value is obtained from the previous one by adding 3, starting over when the King is reached. This type of order is usually attributed to Si Stebbins (late 19th century), but has existed at least since the 16th century:

$$1, 4, 7, 10, K, 3, 6, 9, Q, 2, 5, 8, J.$$

The common difference, being 3, provides a simple rule for preparing the deck, which is presented in Stebbins' book. The suits are separated into four piles, where the top cards are as follows: A♣4♥7♠10♦. In each pile, the value of the cards should be increasing as cards are added on top (with the pile facing up). Thus, below the A♣ should be the K♣, next the Q♣; below the 4♥ should be the 3♥, and so on. This ensures that not only are the values in progression, but the suits are in CHaSeD order.

To prepare the deck, you draw cards in succession from each pile: A♣, then 4♥, then 7♠, 10♦, K♣ (which has since come into view), and so on.

In fact, you can use any number smaller than 13 for the common difference of the arithmetic progression, although three makes possible the method just described for preparation. In any case, the progression will go through all the numerical values until it returns to the first one. This is because the number 13 is prime! Let's consider a progression of common difference 3. Moving k steps corresponds to adding $3 \times k$ to the value of a card. If we get a card with the same value, this means that $3 \times k$ is a multiple of 13. Now, exactly because

13 is a prime, this is never possible for any value of k. The reasoning can be applied to any common difference smaller than 13.

Eight Kings

This is another possible order for the card values, which is associated to a mnemonic: "Eight Kings threatened to save ninety-five Queens for one sick Knave." The order is: 8, K, 3, 10, 2, 7, 9, 5, Q, 4, 1, 6, J.

Positional systems

Any integer greater than 1 can serve as a base for a positional numbering system. Our usual system, the decimal, uses powers of base 10 to express numbers. For example, when we write 23654, we are condensing

$$23654 = 2 \times 10000 + 3 \times 1000 + 6 \times 100 + 5 \times 10 + 4$$
$$= 2 \times 10^4 + 3 \times 10^3 + 6 \times 10^2 + 5 \times 10^1 + 4 \times 10^0$$

In the binary system, 2 plays the role of 10. In this base, 14 is represented as follows:

$$14 = 8 + 4 + 2 = 1 \times 2^3 + 1 \times 2^2 + 1 \times 2^1 + 0 \times 2^0 = 1110_2$$

As we said, we can use any number greater than 1 for a base. In base 3, we have

$$14 = 9 + 3 + 2 = 1 \times 3^2 + 1 \times 3^1 + 2 \times 3^0 = 112_3$$

Therefore, given a base, to each positive integer is associated a sequence of coefficients, as in the expressions above. Note that the coefficients are always smaller than the base: in binary we only have the digits 0 and 1, in base 3 we only have 0, 1, and 2, and in base 10, as we know, we only have digits from 0 to 9.

When we work in base 2, the only coefficients that can appear are 0 or 1. This means that expressing a number in base 2 is the same as expressing it as a sum of powers of 2, which are the powers that are multiplied by the coefficients 1. This method can be used to quickly get the expression of a number in base 2 (as we did with the number 14, above). For example,

$$5 = 4 + 1 = 2^2 + 2^0 \qquad 12 = 8 + 4 = 2^3 + 2^2$$

In base 2, this decomposition corresponds to a sequence of zeros and ones. Thus, in base 2, five is written 101 and twelve is written 1100. The digits 1 appear in the positions corresponding to the powers of 2 that occur in the decomposition:

	$2^3 = 8$	$2^2 = 4$	$2^1 = 2$	$2^0 = 1$
5	0	1	0	1
12	1	1	0	0

Modular arithmetic

We have all learned to add and multiply whole numbers. However, when we do math with hours, for example, we use a different arithmetic. For example, if we go to the movies for a session that starts at 11 p.m. and the movie lasts two hours, we leave the theatre at 1 a.m., which means that in this case, $11 + 2 = 1$.

Even if the previous equality is wrong, there is a way to make it correct: Identify two numbers whose difference is a multiple of 12 (or 24, if you prefer this system). Thus, we write $11 + 2 = 13$ and $13 \equiv 1$ (mod 12) to indicate this identification, reading "13 is congruent with 1, modulo 12".

There is nothing special here with the number 12, it is possible to make these identifications with any number. If we choose a certain natural n, we can define in the integers an equivalence relation, saying that $a \equiv b$ (mod n) if the difference $a - b$ is divisible by n. This relation is called congruence modulo n and determines classes in the integers, which are naturally called congruence classes.

When we divide a number a by n with remainder, it is easy to see that this remainder is congruent with a modulo n: $a = q \times n + r$, so $a - r = qn$ is a multiple of n.

Thus, saying that a and b are congruent modulo n is the same as saying that they have the same remainder when divided by n.

Addition and multiplication can also be done modulo n, by identifying the elements that are in the same class. For example, if $n = 5$,

$$4 + 7 = 11 \equiv 1 \quad (\text{mod } 5) \qquad 4 \times 7 = 28 \equiv 3 \quad (\text{mod } 5)$$

This is because $11 - 1 = 10$ and $28 - 3 = 25$, both multiples of 5.

These operations do not distinguish between numbers of the same class. If we had chosen numbers congruent with 4 and 7, modulo 5, the result of the operations would have been the same. For example, $9 \equiv 4 \ (\text{mod } 5)$, $12 \equiv 7 \ (\text{mod } 5)$ and

$$9 + 12 = 21 \equiv 1 \quad (\text{mod } 5) \qquad 9 \times 12 = 108 \equiv 3 \quad (\text{mod } 5)$$

For $n = 2$, the operations look familiar:

$$0 + 0 \equiv 1 + 1 \equiv 0 \quad (\text{mod } 2) \qquad 0 \times 0 \equiv 0 \times 1 \equiv 0 \quad (\text{mod } 2)$$

$$1 \times 1 \equiv 1 \quad (\text{mod } 2)$$

Since the odd numbers are congruent with 1 and the even numbers are congruent with 0, modulo 2, what we are saying is that the sum of even numbers is even, the sum of odd numbers is even, and the sum of an even with an odd one is odd. For the product, analogous interpretations can be made.

The congruence modulo 9 deserves a little highlighting. It is simple to see that any power of 10 is congruent to 1, modulo 9, since it is written as $99 \ldots 9 + 1$. So, if a number n is written as

$$n = a_0 + a_1 \times 10 + \cdots + a_t \times 10^t$$

as the congruences respect sums and products, we obtain

$$n \equiv a_0 + a_a + \cdots + a_t \quad (\text{mod } 9)$$

This means that when we add up the digits of a number we get another that is congruent with the first modulo 9, that is, that has the same remainder when divided by 9. As a consequence, the difference between a number and the sum of its digits is always a multiple of 9. This is the base for the traditional procedure that goes by the name of *casting out nines*.

Bibliography

Brushwood, B. (n.d.). Scam school, https://www.youtube.com/user/scam school.

de Sequeira, G. C. (1612). *Thesouro de Prudentes de Gaspar Cardozo de Sequeira — Livro 3, Tratado 3*. Edited by Hugo Almeida, Jorge Nuno Silva, Pedro J. Freitas, and Tiago Hirth. (Ludus, 2024).

Diaconis, P. and Graham, R. (2015). *Magical Mathematics: The Mathematical Ideas That Animate Great Magic Tricks* (Princeton University Press), https://books.google.pt/books?id=W26YDwAAQBAJ.

Fulves, K. (2012). *Self-Working Card Tricks*, Dover Magic Books (Dover Publications), https://books.google.pt/books?id=8NtYMqsWitkC.

Gardner, M. (2014). *Mathematics, Magic and Mystery*, Dover Math Games & Puzzles (Dover Publications), https://books.google.pt/books?id=WkS4BQAAQBAJ.

Gardner, M. (2020). *Hexaflexagons and Other Mathematical Diversions: The First Scientific American Book of Mathematical Puzzles and Games* (American Mathematical Soc.), https://books.google.pt/books?id=UE0FEAAAQBAJ.

Grime, J. (n.d.). Singingbanana, https://www.youtube.com/user/singingbanana.

MacTier, A. (2000). *Card Concepts: An Anthology of Numerical & Sequential Principles Within Card Magic* (Lewis Davenport Limited), https://books.google.pt/books?id=VndjHQAACAAJ.

McOwan, P. and Parker, M. (2010). The manual of mathematical magic, https://mathematicalmagic.com/.

Mismag822 (n.d.). Mismag822, https://www.youtube.com/user/mismag822.

Mulcahy, C. (n.d.). Card colm, http://cardcolm-maa.blogspot.com, online column supported by Mathematical Association of America.

Mulcahy, C. (2013). *Mathematical Card Magic: Fifty-Two New Effects*, AK Peters/CRC Recreational Mathematics Series (CRC Press), https://books.google.pt/books?id=CzrSBQAAQBAJ.

Silva, J. N. (2008). *Os Matemágicos Silva*, Bisca lambida (Apenas).

Stebbins, S. (2019). *Card Tricks And the Way They Are Performed* (Independently Published), https://books.google.pt/books?id=faaqyw EACAAJ.

Index

Alex Elmsley, 47, 48, 51, 173
Australian shuffle, 60, 104, 182

base
    binary, 185
    decimal, 185
    ternary, 185
Bell numbers, 168
Bill Nord, 30
binary base, 185
Bob Hummer, 71, 76, 77, 79, 173

casting out nines, 187
CATO shuffle, 183
Charles Hudson, 77
Charles Jordan, 36, 63
Charles S. Peirce, 65
CHaSeD order, 12, 86, 91, 95, 116, 124, 129, 162, 164, 166, 184
Colin Wright, 125
Colm Mulcahy, 127
congruence, 186
cut shuffle, 7, 181

Dai Vernon, 51, 134, 174
de Bruijn's sequence, 171
decimal base, 185
Don Costello, 156
dovetail shuffle, 180, 181
down-and-under shuffle, 182

Eight Kings order, 139, 166, 185

false shuffle, 179
Fan Chung, 168
Faro shuffle, 48, 180
Felix Klein, 4, 5
Fibonacci numbers, 11
Findley, 86
Finnell's Free Cut Principle, 56

Gaspard Monge, 181
Gathering for Gardner, 18
    Europe, 18
Gathering for Gardner Europe, 18
Gilbreath, 135–139, 141, 142, 144, 145
    first princliple, 135–138
    on steroids, 144
    second princliple, 139, 141, 142, 144, 145

Harry Houdini, 174
Henry Christ, 14
Hummer shuffle, 78, 79, 183

in-and-out shuffle, 59, 66, 92, 180
inverse shuffle, 59

Jim Steinmeyer, 27
John Railing, 18
Jorge Nuno Silva, 75, 90, 104

Klondike shuffle, 182

Laura Silva, 1
Lennart Green, 27, 53, 77, 174

linear shift-register, 171
Luís Sequeira, 122
Luísa do Vítor, 74
Luca Pacioli, 104

Manuel Silva, 1
Margarida Pinto, 69
Martin Gardner, 17, 156, 175
milk shuffle, 48, 62, 182
modular arithmetic, 186
Monge shuffle, 59, 60, 67, 181, 182

Norman Gilbreath, 135
numbers
    Bell, 168
    Fibonacci, 11
    prime, 11

order
    CHaSeD, 12, 86, 91, 95, 116,
      124, 129, 162, 164, 166, 184
    Eight Kings, 139, 166, 185
    in suits, 184
    progression, 184
    Sequeira, 162, 163
    Stebbins, 163, 164, 184

Paul Curry, 18
Paul Erdős, 95
Paul LePaul, 173
Pedro Jorge Freitas, 37, 118, 158
Penelope Principle, 48
perfect shuffle, 48
Persi Diaconis, 56, 177
positional writing, 185
prime numbers, 11
progression order, 184

Recreational Mathematics Colloquia,
  177
reverse shuffle, 180

riffle shuffle, 133, 135–137, 139, 141,
  143, 144, 150, 154, 180, 181
Ronald Graham, 56
rosetta shuffle, 143, 181

Sequeira, 161–163
    order, 162, 163
shuffle, 179–181
    Australian, 60, 104, 182
    CATO, 183
    cut, 7, 181
    dovetail, 180, 181
    down-and-under, 182
    false, 179
    Faro, 48, 180
    Hummer, 78, 79, 183
    in-and-out, 59, 66, 92, 180
    inverse, 59
    Klondike, 182
    milk, 48, 62, 182
    Monge, 59, 60, 67, 181, 182
    perfect, 48
    reverse, 180
    riffle, 133, 135–137, 139, 141,
      143, 144, 150, 154, 180, 181
    rosetta, 143, 181
Somerset Maugham, xi
stay-stack, 59
Stebbins, 163–165
    order, 163, 164
Stephen Minch, 173
Steve Belchou, 9
Stewart James, 33
Stewart Judah, 69

ternary base, 185
The Pentacle Club, 173
Tiago Hirth, 128

Winston Churchill, 18

www.ingramcontent.com/pod-product-compliance
Lightning Source LLC
Chambersburg PA
CBHW071740150726
47998CB00005B/1737